T0137348

Springer Theses

Recognizing Outstanding Ph.D. Research

Aims and Scope

The series "Springer Theses" brings together a selection of the very best Ph.D. theses from around the world and across the physical sciences. Nominated and endorsed by two recognized specialists, each published volume has been selected for its scientific excellence and the high impact of its contents for the pertinent field of research. For greater accessibility to non-specialists, the published versions include an extended introduction, as well as a foreword by the student's supervisor explaining the special relevance of the work for the field. As a whole, the series will provide a valuable resource both for newcomers to the research fields described, and for other scientists seeking detailed background information on special questions. Finally, it provides an accredited documentation of the valuable contributions made by today's younger generation of scientists.

Theses are accepted into the series by invited nomination only and must fulfill all of the following criteria

- They must be written in good English.
- The topic should fall within the confines of Chemistry, Physics, Earth Sciences, Engineering and related interdisciplinary fields such as Materials, Nanoscience, Chemical Engineering, Complex Systems and Biophysics.
- The work reported in the thesis must represent a significant scientific advance.
- If the thesis includes previously published material, permission to reproduce this must be gained from the respective copyright holder.
- They must have been examined and passed during the 12 months prior to nomination.
- Each thesis should include a foreword by the supervisor outlining the significance of its content.
- The theses should have a clearly defined structure including an introduction accessible to scientists not expert in that particular field.

More information about this series at http://www.springer.com/series/8790

Anna Siri Luthman

Spectrally Resolved Detector Arrays for Multiplexed Biomedical Fluorescence Imaging

Doctoral Thesis accepted by
the University of Cambridge, Cambridge, UK

Author
Dr. Anna Siri Luthman
Department of Physics
University of Cambridge
Cambridge, UK

Supervisor
Dr. Sarah Bohndiek
Department of Physics
University of Cambridge
Cambridge, UK

ISSN 2190-5053 ISSN 2190-5061 (electronic)
Springer Theses
ISBN 978-3-030-07473-9 ISBN 978-3-319-98255-7 (eBook)
https://doi.org/10.1007/978-3-319-98255-7

This Springer imprint is published by the registered company Springer Nature Switzerland AG
The registered company address is: Gewerbestrasse 11, 6330 Cham, Switzerland

Jag tror på tvivlet.

Tvivlet är all kunskaps grund och all förändrings motor.

...

I believe in doubt.

Doubt is the foundation of all knowledge and the engine of all change.

Tage Danielsson
Tankar från Roten (1974)

To my grandfather Lars Luthman

an engineer, an eternal student and a wonderful human being

Supervisor's Foreword

Optical imaging plays an important role in the early diagnosis of cancer. In particular, the use of optical endoscopes to relay image information from deep within the body to the external observer is widespread. At present, however, the optical information relayed is typically recorded using a standard colour camera, which integrates a Bayer filter array of red, green and blue colour filters to replicate the colour sensing capability of the human eye. This approach restricts the range of wavelengths that can be detected and the number of spectral features that can be resolved. As the early stages of cancer often appear flat, and just a slightly different shade of pink from normal tissue, white light endoscopic imaging using standard colour cameras, provides limited contrast for early detection and hence high miss rates during endoscopic surveillance and screening.

The interactions of light with tissue go far beyond the simple red, green and blue. In addition to those interactions that occur intrinsically with different biomolecules in the tissue, it is also possible to use contrast agents that can specifically enhance the contrast between healthy and diseased tissues. These contrast agents may be untargeted and give rise to contrast due to differences in tissue structure or vascularisation, or they may be targeted to specific cell surface receptors or other biological processes that are known to change during disease progression. The detection of these contrast agents in tissue can often be difficult due to the high background signals that arise from the intrinsic tissue interactions. Furthermore, the number of contrast agents that can be detected simultaneously is limited to just two or three, due to the inability to resolve their spectral signatures.

In this thesis, Siri Luthman describes the design, development, characterisation and application of two novel fluorescence instruments to address these challenges and enable imaging of multiple fluorescent contrast agents simultaneously. These instruments are based on recent advances in spectrally resolved detector arrays (SRDAs). The standard colour camera is a simple SRDA containing only three colour channels. The novel SRDAs used in this thesis contain many more colours, ranging from 16 towards 100 colour channels. Using these compact and robust detectors, with potential for low-cost manufacture, in biomedical applications enables new avenues of exploration that were not possible before.

As with any study using a new technology, a careful technical characterisation of the technology was first required and the first chapter of the thesis is dedicated to developing a detailed understanding of the strengths and weaknesses of the technology. A comprehensive assessment of the SRDA quantum efficiencies for a range of different filter implementations was performed. In particular, Siri paid attention to the angular acceptance of the filters used in the SRDAs and how F number affects the spectral resolution.

With a detailed understanding of the SRDA performance, Siri then went on to apply these novel cameras in two different applications, namely in a wide-field fluorescence imaging system and in an endoscopic fluorescence imaging system. The work presented in this thesis shows for the first time that not only can this new type of camera be used for fluorescence imaging, but that it is able to resolve signals from up to seven different dyes simultaneously, a level of multiplexing not previously achieved. Furthermore, her bimodal endoscope was able to perform both reflectance and fluorescence spectral imaging using these cameras; this is the first example of such a device, and the potential for future application in gastrointestinal imaging was demonstrated in an ex vivo pig oesophagus model.

With her dissertation, Siri made an important contribution towards understanding how this new SRDA technology can be applied in biomedicine. As highlighting the exciting potential of SRDAs in fluorescence imaging, she also identified the need for improved optical throughput and greater refinement in the peak wavelength and bandwidth of the available filters. These are active areas of ongoing research in my own laboratory and others around the world, illustrating how Siri's work has opened up a new research direction in the field of biomedical imaging.

Cambridge, UK
July 2018

Dr. Sarah Bohndiek

Abstract

The ability to resolve multiple fluorescent emissions from different biological targets in video rate applications, such as endoscopy and intraoperative imaging, has traditionally been limited by the use of filter-based imaging systems. Hyper- and multispectral imaging facilitate the detection of both spatial and spectral information in a single data acquisition; however, instrumentation for spatiospectral data acquisition is typically complex, bulky and expensive. This thesis seeks to overcome these limitations by using recently commercialised compact and robust hyper-/multispectral cameras based on spectrally resolved detector arrays.

Following sensor calibrations, which devoted particular attention to the angular sensitivity of the sensors, we integrated spectrally resolved detector arrays into a wide-field and an endoscopic imaging platform. This allowed multiplexed reflectance and fluorescence imaging with spectrally resolved detector array technology in vitro, in tissue-mimicking phantoms, in an ex vivo oesophageal model and in vivo in a mouse model. A hyperspectral linescan sensor was first integrated into a wide-field near-infrared reflectance-based imaging set-up to assess the suitability of spectrally resolved detector arrays for in vivo imaging of exogenous fluorescent contrast agents. Using this fluorescence hyperspectral imaging system, we could accurately resolve the presence and concentration of seven fluorescent dyes in solution. We also demonstrated high spectral unmixing precision, signal linearity with dye concentration, at depth in tissue-mimicking phantoms and delineation of four fluorescent dyes in vivo. After the successful demonstration of multiplexed fluorescence imaging in a wide-field set-up, we proceeded to combine near-infrared multiplexed fluorescence imaging with visible light spectral reflectance imaging in an endoscopic set-up. A multispectral endoscopic imaging system, capable of simultaneous reflectance and fluorescence imaging, was developed around two snapshot spectrally resolved detector arrays. In the process of system integration and characterisation, methods to characterise and predict the imaging performance of spectral endoscopes were developed. With the endoscope, we demonstrated simultaneous imaging and spectral unmixing of chemically oxy/deoxygenated blood and three fluorescent dyes in a tissue-mimicking phantom, and of two fluorescent dyes in an ex vivo oesophageal porcine model. With further developments, this technology has the potential to become applicable in medical imaging for detection of diseases such as gastrointestinal cancers.

Parts of this thesis have been published in the following articles:

Publications

A S Luthman, S Dumitru, I Quiros-Gozalez, J Joseph, S E Bohndiek (2017), Fluorescence hyperspectral imaging (fHSI) using a spectrally resolved detector array, J Biophotonics 10(6–7) pp. 840–853

T Sawyer, **A S Luthman**, S E Bohndiek (2017), Evaluation of illumination system uniformity for wide-field biomedical hyperspectral imaging, J Opt 19(3) pp. 1–10

A S Luthman, D J Waterhouse, L Bollepalli, G S D Gordon, J Joseph, S E Bohndiek (2018), Bimodal Multispectral Endoscope based on Spectrally Resolved Detector Arrays. Under Consideration for Publication

J Yoon, J Joseph, G S D Gordon, **A S Luthman**, D J Waterhouse, C Williams and S E Bohndiek (2018), Development of a clinically translatable hyperpectral endoscope using a line-scanning spectrograph. Under Consideration for Publication

D J Waterhouse, **A S Luthman**, J Yoon, G S D Gordon and S E Bohndiek (2018), Quantitative evaluation of comb-structure removal methods in multispectral fiberscopic imaging. Under Consideration for Publication

Conference Proceedings

A S Luthman, S E Bohndiek (2015), Experimental evaluation of a hyperspectral imager for near-infrared fluorescent contrast agent studies, Proc SPIE Int Soc Opt Eng Vol. 9318, 93180H

A S Luthman, S Dumitru, I Quirós-Gonzalez, S E Bohndiek (2016), Hyperspectral fluorescence imaging with multi wavelength LED excitation, Proc SPIE Int Soc Opt Eng Vol. 9711, 971111

D J Waterhouse, **A S Luthman**, S E Bohndiek (2017), Spectral band optimization for multispectral fluorescence imaging, Proc SPIE Int Soc Opt Eng Vol. 10057, 1005709

T W Sawyer, **A S Luthman**, S E Bohndiek (2017), Evaluation of illumination systems for wide-field hyperspectral imaging in biomedical applications, Proc SPIE Int Soc Opt Eng Vol. 10068, 1006818

A S Luthman, D J Waterhouse, L Bollepalli, J Joseph, S E Bohndiek (2017), A multispectral endoscope based on spectrally resolved detector arrays, Proc SPIE Int Soc Opt Eng Vol. 10411, 104110A

Acknowledgements

Firstly, I would like to acknowledge the incredible support, scientific guidance and inspiration provided by my supervisor Dr. Sarah Bohndiek. I would especially like to thank her for having the trust to allow me to join the newly founded VISION Lab group in 2013. I have developed both professionally and personally by taking part in the establishment of a new research group.

I would also like to acknowledge the scientific training and research inspiration I have received through work and discussions with colleagues in the VISION Lab group at the University of Cambridge, especially Dale Waterhouse, Dr. James Joseph, Dr. George Gordon, Dr. Isabel Quiros-Gonzalez, Dr. Laura Bollepalli, Dr. Massimiliano di Pietro, Dr. Calum Willimas, Sebastian Dumitru and Travis Sawyer, some of which have also contributed directly to work presented in this thesis; their help has been much appreciated, and their relative contributions appropriately acknowledged in the main text. I would also like to thank the mechanics and electronics workshops at the Cavendish Laboratory, Dr. Renato Turchetta and Ben Marsh of Rutherford Appleton Laboratories for access to their sensor test system, the hyperspectral imaging groups at Imec and SILIOS for swiftly answering my technical queries, and Prof. Lise Lyngsnes Randeberg at NTNU for helpful scientific discussions. Thanks also to my thesis assessors for their future feedback and comments.

I would additionally like to acknowledge the funding from the Engineering and Physical Research Council, the Schiff Foundation, and the Foundation Blanceflor without whose financial support this project would not have been possible. This work was also supported by the Engineering and Physical Sciences Research Council.

Last but not least, I would like to express my great gratitude to friends and family who have supported me throughout my studies. To my mother and father who encourage me to be ambitious and pursue my passions, to my sister Iris who provides a wise perspective, and lastly, but most importantly, to my fiancé Nick Farley who keeps reminding me that the world extends far beyond the walls of the Cavendish Laboratory; thank you for sharing valleys and peaks.

Contents

Acronyms/Abbreviations

5-ALA	5-Aminolevulinic acid
ADC	Analogue-to-digital converter
AF	AlexaFluor
AF	Autofluorescence
AFI	Autofluorescence Imaging
AOI	Angle of incidence
AOTF	Acousto-optical tunable filter
ASGE	American Society for Gastrointestinal Endoscopy
AUC	Area under curve
CASSI	Coded aperture snapshot spectral image
CCD	Charge-coupled device
CFB	Coherent fibre bundle
CMOS	Complimentary metal-oxide semiconductor
CNR	Contrast-to-noise ratio
CRA	Chief ray angle
CTIS	Computed tomography imaging spectroscopy
Cy	Cyanine
DN	Digital number
DR	Dynamic range
DSNU	Dark signal non-uniformities
EAC	Oesophageal adenocarcinoma
EMA	European Medicines Agency
EMCCD	Electro-multiplying charge-coupled device
EMVA	European Machine Vision Association
F/#	F number
FDA	Federal Drug Administration
fHSI	Fluorescence hyperspectral imaging
FOV	Field of view
FPA	Focal-plane array
FPGA	Field programmable gate array

FPN	Fixed-pattern noise
FWHM	Full-width half-maximum
GI	Gastrointestinal
GPU	Graphical processing unit
GUI	Graphical user interface
HGD	High-grade dysplasia
HSI	Hyperspectral imaging
IMS	Image mapping spectrometry
IRIS	Image-replicating imaging spectrometry
LCD	Liquid-crystal display
LCTF	Liquid-crystal tunable filter
LED	Light-emitting diode
LGD	Low-grade dysplasia
LS	Least squares
LUT	Look-up table
MEMS	Microelectromechanical systems
MFD	Mode field diameter
MSE	Mean squared error
MSI	Multispectral imaging
NA	Numerical aperture
NAD	Nicotinamide adenine dinucleotide
NBI	Narrowband imaging
ND	Neutral density
NIR	Near infrared
NNLS	Non-negative least squares
OD	Optical density
OPD	Optical path difference
OSP	Orthogonal subspace projection
PBS	Phosphate buffer solution
PCA	Principal component analysis
PpIX	Protoporphyrin IX
PRNU	Photo response non-uniformities
PTC	Photon transfer curve
QE	Quantum efficiency
RMSE	Root-mean-square error
ROI	Region of interest
SBR	Signal-to-background ratio
S-Cy	Sulfo-Cyanine
SFE	Scanning fibre endoscope
SHIFT	Snapshot hyperspectral imaging Fourier transform
SNR	Signal-to-noise ratio
SRDA	Spectrally resolved detector array
SUP	Spectral unmixing precision

USAF	United States Air Force
VHG	Volume holographic grating
WD	Working distance
WLE	White light endoscopy

Chapter 1
Summary and Rationale

The way light is scattered, transmitted, absorbed and readmitted as it passes through matter contains information about the spatial and chemical composition of an object. Biomedical researchers and clinicians take advantage of this to study biological processes, to diagnose and to monitor diseases. As disease progresses, the morphology and chemical constituents of tissue change; these changes can be recorded in the absorption, reflectance, or fluorescence spectra of the tissue. The applications of light in clinical imaging are, however, often limited to either taking a wide-field picture, or a spectral measurement at a single point. This limits the use of spectral data in real-time clinical intraoperative and diagnostic imaging applications.

Acquisition of spectral data allows diagnostic parameters to be extracted. When spectral data are combined with wide-field images, these diagnostic parameters may be overlaid on wide-field clinical images to aid diagnosis and guide treatment. The use of spectral image data to probe light interactions with tissue (Fig. 1.1a) is therefore an active area of preclinical and clinical research. For example, the differential spectral absorption of oxy/deoxygenated haemoglobin has been shown to enable read-out of the oxygenation status of tissue to aid diagnosis and guide intraoperative decisions [1–3]. Tissue fluorescence—the absorption of light at one wavelength followed by the subsequent emission of light at a usually longer wavelength—is frequently also used for imaging [4–6]. The endogenous fluorophores in healthy and diseased tissue can differ. Tissue autofluorescence (AF) can therefore reveal a difference between healthy and diseased tissue [4]. Intrinsic fluorescent properties of tissue arising from molecules, such as oxygenated/reduced nicotinamide adenine dinucleotide (NAD) and collagen, can for example be targeted to aid the diagnosis of cancer during endoscopy [4].

It can be difficult to extract specific clinical information from intrinsic chromophores due to the often unknown molecular composition and uncertainty in the contribution of each fluorophore to the overall spectral signal [4]. Extrinsically

A. S. Luthman, *Spectrally Resolved Detector Arrays for Multiplexed Biomedical Fluorescence Imaging*, Springer Theses,
https://doi.org/10.1007/978-3-319-98255-7_1

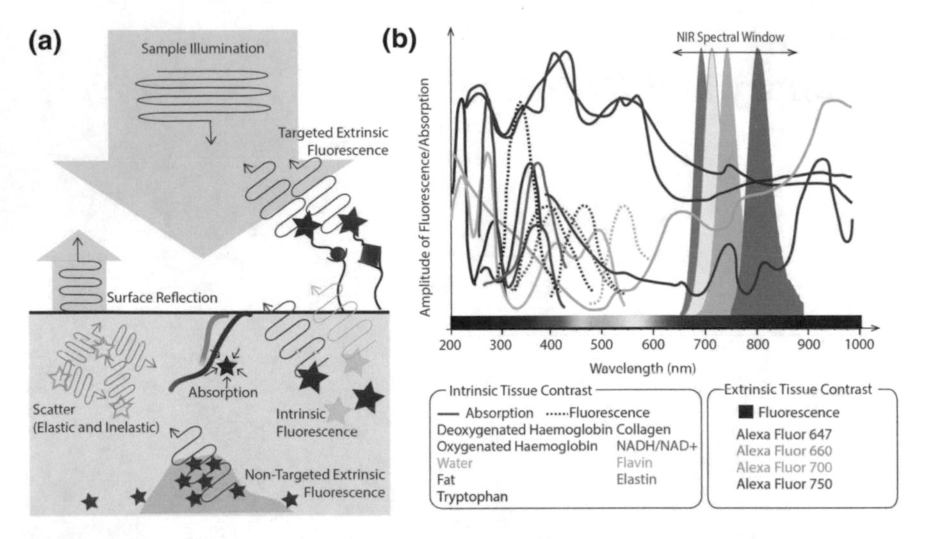

Fig. 1.1 **a** Intrinsic tissue contrast is accessible from the reflectance (light blue arrow), scatter (light blue stars) and the absorption (black star) of light as well as from the AF from intrinsic fluorophores (dark purple and green stars). Tissue contrast can also be enhanced by the administration of extrinsic fluorescent contrast agents. Extrinsic contrast agents may either be non-targeted, to provide increased fluorescence contrast via organ retention (pink stars), or targeted towards specific cell surface receptors (light and dark red stars). **b** The spectral absorption and fluorescence of intrinsic chromophores can be used to extract diagnostic information. In the NIR spectral region, the scatter, absorption and AF of tissue decrease. This yields the 'NIR spectral window' in which extrinsic fluorophores may be imaged with high SBRs. For illustrative purposes the fluorescence spectra of four NIR fluorescent dyes, used in this thesis, are shown in this spectral region. Figure **b** is modified from www.biomedima.org [7] and fluorescent dye spectra were obtained from the supplier (Invitrogen)

administered fluorescent contrast agents are therefore used to increase the intensity differences in the images. Non-targeted fluorescent contrast agents, such as methylene blue, can for example provide increased tissue contrast (i.e. a higher relative intensity) via differential organ retention [6]. Targeted extrinsic molecular probes may also be used to perform optical molecular imaging. An optical molecular probe typically consists of an optical reporter, such as a fluorescent molecule conjugated to a targeting part, which binds to a specific cell surface target [8]. Extrinsic fluorescence signals can therefore allow direct and specific visualisation of molecular mechanisms [8]. Although still an emerging clinical imaging technique, optical molecular imaging has already shown the potential to improve imaging performance in several clinical applications, such as the diagnostic performance in a range of gastrointestinal (GI) cancers [9–11]. These studies have typically been limited to measuring the fluorescence signal from a single molecular probe [9, 11]. True spectral imaging may be used to simultaneously acquire tissue information from multiple intrinsic and extrinsic fluorophores; here referred to as multiplexed biomedical fluorescence imaging. Multiplexed imaging may help overcome one of the main challenges of

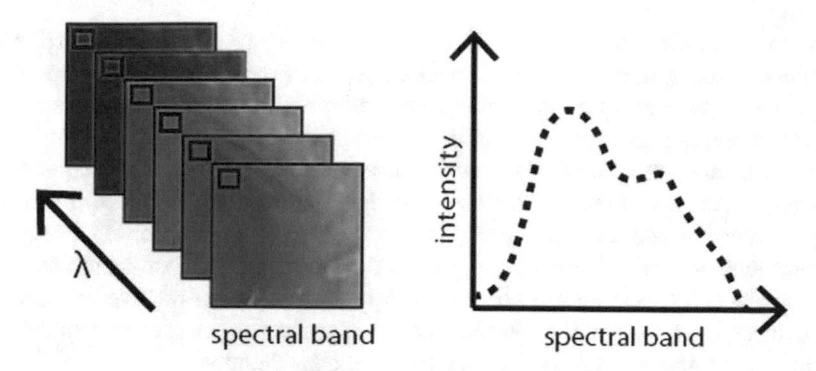

Fig. 1.2 HSI/MSI systems acquire spatial image data in several spectral bands, such that spectral information can be extracted from each spatial pixel

biomedical fluorescence imaging: to extract diagnostic information from diverse patient populations with varying levels of biomarker expression [8, 10].

When sampling a broad spectral range, the spectral acquisition of intrinsic and extrinsic fluorescence may also be matched to spectral regions of high or low tissue AF (Fig. 1.1b). The relatively stronger AF in the visible spectral region decreases when moving towards the near-infrared (NIR) wavelengths [6]. The lower AF gives rise to the 'NIR window' [12, 13], in which extrinsic fluorophores may be imaged with higher signal-to-background ratios (SBRs) [12]. In comparison to the visible spectral region, tissue absorption and scattering is also relatively low within the NIR [6, 12, 13]. This increases the penetration depth of light in tissue and allows imaging of extrinsic fluorescent contrast agents with high SBRs [12]. Hence, while it is thought more suitable to extract intrinsic tissue contrast from the visible spectral region, this thesis therefore focuses on technology evaluation for performing multiplexed imaging of extrinsic fluorophores in the NIR.

Simultaneous measurements of multiplexed intrinsic and extrinsically administered biomarkers could improve medical imaging. The instrumentation challenges associated with the acquisition of spectral image data have, however, limited advances in biomedical multiplexed imaging, as instrumentation tends to be bulky and expensive. Hyperspectral and multispectral imaging (HSI/MSI) systems acquire spectral image data (Fig. 1.2). HSI/MSI involve the acquisition of image data at several wavelengths (spectral bands), such that the spectrum of each image pixel may be extracted. Whereas MSI sensors sample the spectral region of interest sparsely by acquiring measurements in ~10 separate spectral bands, HSI sensors acquire data in ~100 or more spectral bands to provide contiguous spectral sampling [14].

Multiplexed biomedical fluorescence imaging is achieved by separating the reflectance and/or fluorescence signal from intrinsic and/or extrinsic biomarkers using the spectral information contained in the MSI/HSI data cube. To derive useful information, it is necessary to relate the spectral data to the chemical composition of the sample. The relationships between the spectra and the chemical composition can be established via spectral unmixing. The end goal of spectral unmixing is to extract the abundances of the individual components in a sample by separating the reflectance, absorption or fluorescence spectra of its constituents. The spectra of the individual components are referred to as endmembers, and spectral unmixing can involve estimating the number of endmembers, their spectral signatures and their abundances at each spectral pixel [15]. Since the absorption, reflectance and fluorescence spectra of many intrinsic and extrinsic biomarkers are often previously known (or can be acquired via a straightforward reference measurement) the unmixing problem can typically be reduced to an inversion problem, with aim to estimate the abundance of each endmember. To accurately obtain the relative endmember abundances, the impact of the instrument on the spectra and the spectral dependency of the instrument response need to be taken into consideration.

Extraction of 3-D data (x–y–λ) from typical 2-D sensors (x–y) requires translation of the spectral dimension of the 3-D data cube to either a spatial or temporal data dimension. While several methods allow extraction of the spectral dimension, many approaches fail to meet the constraints on compactness, robustness and cost effectiveness required for efficient integration into clinical imaging applications. Spectrally resolved detector arrays (SRDAs), which integrate spectral filters in front of or directly on top of the image sensor, have the potential to meet these requirements. This thesis evaluates the potential of the first commercially available SRDAs (manufactured by Imec and SILIOS) for multiplexed fluorescence imaging in biomedical applications. Two types of SRDAs were investigated: snapshot and linescan SRDAs.

The concept behind snapshot SRDAs is very similar to that behind traditional RGB cameras: spectral filters are deposited in a regular pattern on an underlying image sensor. For linescan SRDAs, the spectral filters are deposited row-wise across the sensor allowing HSI data to be acquired by spatially scanning the scene. With manufacturing developments, snapshot sensors based on the same technique also became commercially available. These snapshot sensors are based on the regular deposition of spectral filters in a mosaic pattern, forming spectral macro-pixels, such that spectral data can be acquired at video-rate. MSI data cube are generated post-data acquisition by interpolating between pixels in adjacent macro-pixels.

Several approaches to realising SRDAs have been proposed. Spectral filtering via pixel-level deposition of Fabry–Pérot interferometers on an underlying image sensor is, however, the most mature technology. The maturity of Fabry–Pérot technology makes it the most promising approach to achieve rapid integration of SRDAs into biomedical and clinical imaging applications. This thesis therefore evaluates SRDAs realised via Fabry–Pérot technology.

Briefly, Fabry–Pérot filters form high transmission [16] and narrow bandpass spectral filters [17], by separating two reflecting surfaces by a spacer layer [17]. Light in resonance within the spacer layer will constructively interfere in the cavity and pass through the filter, whilst out-of-resonance light will de-constructively interfere and be blocked [18]. The spectral selectivity of the filters are therefore determined by the distance propagated between subsequent reflections. Since a shallow angle-of-incidence (AOI) of light results in greater propagation distances between subsequent reflections, the spectral selectivity of a Fabry–Pérot filter is thus highly dependent on the AOI of the light [18]. Characterization of the spectral response of SRDA based on Fabry–Pérot technology should therefore take into account the angle-of-incidence. This may be done by controlling the aperture F Number (F/#) of the camera objective used with the SRDA, where decreasing the F/# increases the AOI [19]. In addition to standard camera calibrations, we therefore also made significant efforts to characterize the spectral response of the SRDAs at a range of F/#s.

Following calibration and characterisation of the commercial SRDAs, their potential in two biomedical imaging applications were studied; wide-field fluorescence HSI (fHSI) and endoscopic MSI. Calibration methods, instrumentation solutions and data analysis protocols were developed to integrate the SRDAs into imaging systems; these methods may also enable further future use of SRDAs in biomedical and clinical imaging applications.

Each chapter in this thesis contains a literature review which explores prior work and further details the rationale behind the specific experiments presented in each chapter. The presentation of the experimental methods and results is followed by a discussion on the significance of the results and the identification of areas of further work. The main findings of this thesis are subsequently summarised (Chap. 5).
The following is an overview of the chapters of the thesis:

Chapter 2: **Spectral Imaging Systems and Sensor Characterisations** starts with a broad overview of previously reported spectral imaging systems with a special focus on SRDAs and their potential for biomedical and clinical application. For the most effective system integration of SRDAs, the spectral transmission characteristics of the sensor type have to be fully understood. Extensive sensor characterisation was therefore performed. The spectral response of SRDAs is typically highly sensitive to the angle of the incident light. Since relatively limited experimental studies exist on the angular sensitivity of SRDAs, special attention was devoted to the F/# dependence of the SRDAs' spectral response characteristics.

This chapter details the verification of supplier provided quantum efficiency (QE) calibration data of SRDAs commercially manufactured by Imec and SILIOS. A monochromator-integrating sphere system incorporating a variable F/# objective lens was additionally developed to study the angular sensitivity of the SRDAs. In general, it was found that the spectral filter bands follow the trends predicted by theory of light interaction in a Fabry–Pérot cavity [17, 20, 21]; as the F/# of the objective lens decreases, the full-width-half-maximum (FWHM) of the spectral response of the filters broadens, the optical throughput increases and the peak wavelength response is blue shifted.

Chapter 3: **Wide-Field fHSI with a Linescan SRDA** details the integration of a linescan SRDA into a wide-field NIR fluorescence HSI (fHSI) set-up. The wide-field set up was used to demonstrate multiplexed fluorescence imaging of seven fluorescent dyes in vitro and four dyes in vivo in a mouse model. Multiplexed fluorescence imaging with high spectral unmixing precision, signal linearity with dye concentration and at depth in tissue mimicking phantoms were also demonstrated. The fHSI approach presented in this chapter, including background removal, could be directly generalised to broader spectral ranges; for example to resolve extrinsic fluorescence signals from strong tissue reflectance and/or AF signals.

Additionally, whilst the wide-field system has intrinsic value as a small animal pre-clinical imaging platform, it also allowed us to explore the implementation challenges of integrating SRDAs in clinical instrumentation, without adding all the complexities that would be associated with a fully functional endoscopic, or intraoperative, system.

Chapter 4: **A Multispectral Endoscope Based on SRDAs** describes the integration of two snapshot SRDAs into an endoscope. The chapter starts with a rationale for multiplexed endoscopic imaging, which is focused on the potential of MSI endoscopy to improve cancer detection during endoscopic screening of the gastrointestinal (GI) tract. A survey of the literature shows that multiplexed reflectance and fluorescence imaging have the potential to improve the detection rate of GI cancers. A bimodal multispectral endoscope was therefore developed to combine spectral imaging of white light reflectance, providing contrast based on the absorption of endogenous chromophores, and molecular biomarkers, using exogenous fluorescent contrast agents.

This chapter details the design, assembly, technical characterisation and the initial imaging performance evaluation of the endoscope in a tissue mimicking phantom and an ex vivo oesophageal porcine model. The preliminary imaging results showed that the endoscope is capable of concurrent imaging and spectral unmixing of oxy/deoxygenated blood and three fluorescent dyes. With further technical development, this technology shows promise to improve the currently low early detection rates of cancer during endoscopic screening of the GI tract.

Chapter 5: **Outlook and Overall Conclusions** summarises the main findings of this thesis and details additional work required prior to clinical use of SRDAs for multiplexed biomedical fluorescence imaging.

It was shown that SRDAs provide a compact, robust and also potentially cost effective approach to multiplexed biomedical fluorescence imaging. Following sensor characterisation, SRDAs were integrated into a wide-field reflectance and an endoscopic fluorescence imaging platform. This allowed us to successfully demonstrate multiplexed reflectance and fluorescence imaging with SRDA technology in vitro, in tissue mimicking phantoms, in an ex vivo oesophageal model and in vivo in a mouse model.

The low fluorescence detection efficiency is currently the main challenge to realise clinical multiplexed biomedical fluorescence imaging with SRDAs. The sensors of the commercial SRDAs used in this thesis are currently not optimised for low light applications, such as fluorescence imaging. Detection efficiencies may therefore in the future be improved by depositing the spectral filters on scientific grade sensors and through application optimised spectral filter designs. With these future technological developments, SRDAs promise to provide a robust and compact approach to biomedical and clinical multiplexed fluorescence imaging for improved diagnoses and image guided treatments.

References

1. Y. Fawzy, S. Lam, H. Zeng, Rapid multispectral endoscopic imaging system for near real-time mapping of the mucosa blood supply in the lung. Biomed. Opt. Express **6**(8), 2980–2990 (2015)
2. N.T. Clancy et al., Multispectral imaging of organ viability during uterine transplantation surgery in rabbits and sheep. J. Biomed. Opt. **21**(10), 106006 (2016)
3. J. Pichette et al., Intraoperative video-rate hemodynamic response assessment in human cortex using snapshot hyperspectral optical imaging. Neurophotonics **3**(4), 045003 (2016)
4. S. Coda, P.D. Siersema, G.W.H. Stamp, A.V. Thillainayagam, Biophotonic endoscopy: a review of clinical research techniques for optical imaging and sensing of early gastrointestinal cancer. Endosc. Int. Open **03**, E380–E392 (2015)
5. S.H. Yun, S.J.J. Kwok, Light in diagnosis, therapy and surgery. Nat. Biomed. Eng. **1**(0008) (2017)
6. A.L. Vahrmeijer, M. Hutteman, J.R. van de Vorst, C.J.H. van de Velde, J.V. Frangioni, Image-guided cancer surgery using near-infrared fluorescence. Nat. Rev. Clin. Oncol. **10**, 507–518 (2013)
7. BioMedIma, The near infrared (NIR) optical window (2017), http://www.biomedima.org/?modality=5&slide=362. Accessed 15 Aug 2017
8. M.B. Strum, T.D. Wang, Emerging optical methods for surveillance of Barrett's oesophagus. Gut **64**, 1816–1823 (2015)
9. B.P. Joshi et al., Multimodal endoscope can quantify wide-field fluorescence detection of Barrett's neoplasia. Endoscopy **48**(2), A1–A13 (2015)
10. J. Hoon Lee, T.D. Wang, Molecular endoscopy for targeted imaging in the digestive tract. Lancet Gastroenterol. Hepatol. **1**(2), 147–155 (2016)
11. J. Burggraaf et al., Detection of colorectal polyps in humans using an intravenously administered fluorescent peptide targeted against c-Met. Nat. Med. **21**(8), 955–966 (2015)
12. G. Hong, A.L. Antaris, H. Dai, Near-infrared fluorophores for biomedical imaging. Nat. Biomed. Eng. **1**(0010), 1–22 (2017)
13. S.L. Jacques, Optical properties of biological tissues: a review. Phys. Med. Biol. **58**, 5007–5008 (2013)
14. G. Lu, B. Fei, Medical hyperspectral imaging: a review. J. Biomed. Opt. **19**(1), 010901 (2014)
15. J.M. Bioucas-Dias et al., Hyperspectral unmixing overview: geometrical, statistical, and sparse regression-based approaches. IEEE J-STARS **5**(2), 354–378 (2012)
16. A. Lambrechts et al., A CMOS-compatible, integrated approach to hyper- and multispectral imaging. in *2014 IEEE International Electron Devices Meeting, IEDM14* (2014), pp. 261–264
17. H.A. Macleod, *Chapter 6: Edge filters and Chapter 7: Band-pass filters, Thin Film Optical Filters*, 3rd edn. (IoP, Bristol, 2000)
18. L. Frey, L. Masarotto, M. Armand, M.L. Charles, O. Lartigue, Multispectral interference filter arrays with compensation of angular dependence or extended spectral range. Opt. Express **23**(9), 11799–11812 (2015)
19. S.P. Burgos, S. Yokogawa, H.A. Atwater, Color imaging via nearest neighbor hole coupling in plasmonic color filters integrated onto a complementary metal-oxide semiconductor image sensor. ACS Nano **7**(11), 10038–10047 (2013)
20. M. Jayapala et al., Monolithic integration of flexible spectral filters with CMOS image sensors at wafer level for low cost hyperspectral imaging in international image sensor workshop, in *Snowbird* (2013)
21. P. Agrawal et al., Characterization of VNIR hyperspectral sensors with monolithically integrated optical filters, in *Proceedings of IS&T International Symposium on Electronic Imaging 2016* (2016)

Chapter 2
Spectral Imaging Systems and Sensor Characterisations

The first HSI/MSI systems were originally developed by NASA [1] to allow astronomers to gain additional information when gazing at the stars. Since then, the application has turned from stars to cells. This has showed that spectrally resolved data have potential to yield higher sensitivity and specificity also for biomedical and clinical imaging applications.

The instrumentation challenges associated with acquisition of spectrally resolved image data have, however, limited the biomedical applications of HSI/MSI systems. Extraction of 3-D data (x–y–λ) from typically 2-D sensors (x–y) requires translation of the spectral data to a spatial or a temporal dimension. While several methods can be used to acquire spectral image data, many approaches fail to meet the constraints on the compactness, robustness and cost effectiveness of instrumentation required for widespread integration into biomedical and clinical imaging applications. SRDAs, which integrate spectral filters in front or on top of standard image sensors, have the potential to meet these requirements. The work presented in this thesis therefore evaluates some of the first commercially available SRDAs for multiplexed biomedical fluorescence imaging.

Following an overview of HSI/MSI systems, with a particular focus on the main strengths and weaknesses of SRDAs, this chapter details the calibration and characterisation of commercial SRDAs manufactured by SILIOS and Imec. The spectral transmission characteristics of the SRDAs need to be fully understood prior to system integration. We have therefore characterised the spectral response of the Imec HSI linescan sensor, the Imec visible and NIR MSI snapshot sensors and the SILIOS MSI snapshot sensor. Since relatively limited experimental studies exist on the angular sensitivity of SRDAs, particular attention was devoted to the dependence of the F/# of the objective lens on the spectral response characteristics of the sensors. In later chapters the integration of the Imec linescan sensor into a wide-field fHSI system (Chap. 3) and the integration of the Imec snapshot sensors into a MSI

A. S. Luthman, *Spectrally Resolved Detector Arrays for Multiplexed Biomedical Fluorescence Imaging*, Springer Theses,
https://doi.org/10.1007/978-3-319-98255-7_2

endoscope (Chap. 4) are presented. Due to time constraints, further work with the SILIOS sensor is not presented in this thesis. The SILIOS sensor is, however, currently being investigated for use in MSI endoscopy [2].

2.1 Literature Review

2.1.1 Overview of MSI/HSI Systems

Metrics for Spectral Imaging System Comparisons

Existing spectral imaging systems are broadly divided into MSI and HSI systems (collectively referred to as spectral imaging systems in this thesis). Whereas MSI systems sparsely sample the spectral region of interest by acquiring measurements in ~10 separate spectral bands, HSI systems provide contiguous spectral sampling in ~100 or more spectral bands [1]. Within these broad categories, the systems are typically further categorised according to their method of data acquisition. Much due to the multidisciplinary nature of the developments of spectral imaging systems and their applications, no consistent use of terminology exists in the literature [3]. In this literature review we have however chosen to base our discussion around the sensor categories presented in the review by Hagen and Kudenov (2013) [3]. Hagen and Kudenov's terminology was chosen as it appears to be relatively widely accepted in the literature. We have, however, expanded the categories defined by Hagen and Kudenov to also incorporate image systems not included in their now somewhat outdated review.

The spectral imaging systems have been compared according to the following metrics; their spatial, spectral and temporal resolution, their optical throughput, their compactness/portability, the computational intensity of data acquisition and analysis, and the potential for cost effective system manufacturing (Table 2.1). The acquisition of 3-D spectral data cubes from 2-D sensors tends to require trade-offs between critical imaging parameters [1]; ideally the imaging system should therefore be selected based on the requirements of the specific imaging application. An attempt has therefore been made to highlight the trade-offs encountered with different imaging system to evaluate their suitability for biomedical and clinical imaging applications.

Table 2.1 further indicates whether an imaging system is currently commercially available or can be easily assembled from off-the-shelf components without extensive system calibrations. Evidently, commercial availability is often a determining factor as to whether an imaging system has been widely applied in biomedical imaging applications. Therefore, when comparing different spectral imaging systems the ease and cost effectiveness of manufacturing need to be considered. Without potential for commercialisation and cost effective manufacturing an imaging system will have limited potential for widespread use in biomedical and clinical imaging applications.

Table 2.1 Overview of the main categories of previously reported HSI/MSI systems. The table summarises the main strengths, weaknesses and trade-offs of each type of spectral imaging system. References to specific studies may be found in the main text

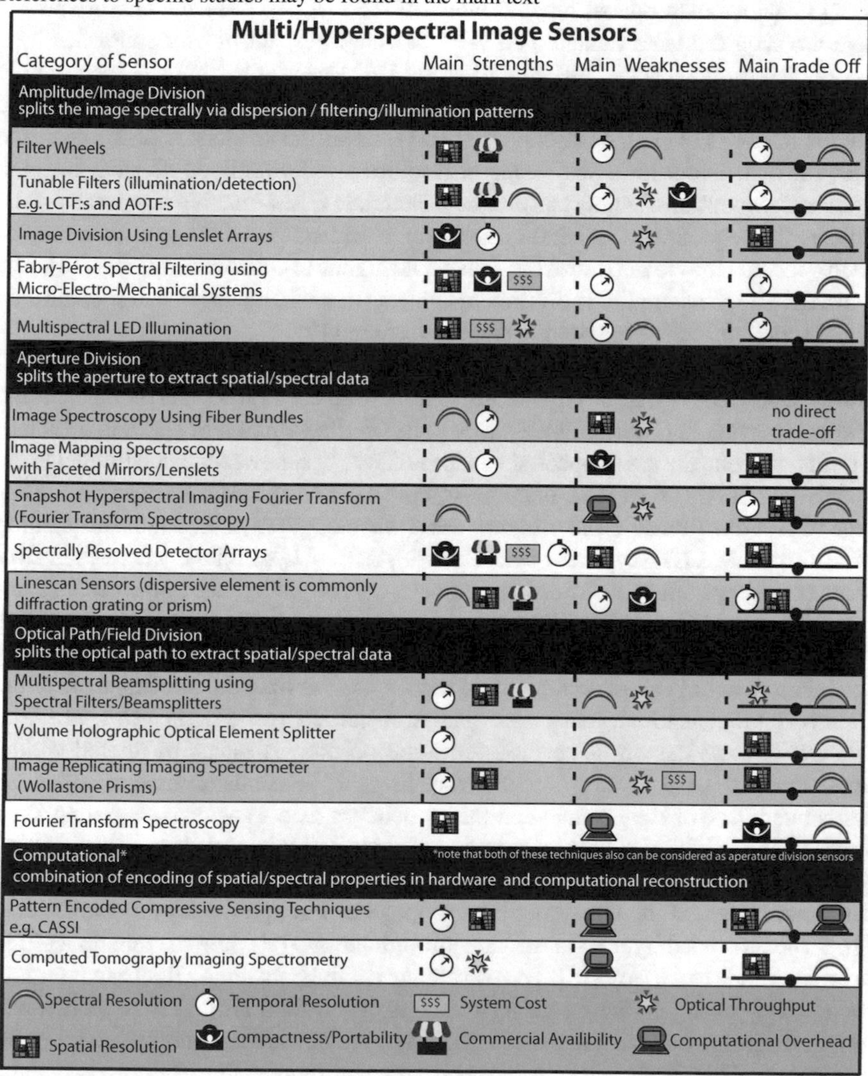

Amplitude/Image Division Spectral Imaging Systems

Amplitude division spectral imaging systems acquire spectral data by spectrally splitting the image via dispersion or filtering on the illumination, or detection, side of the imaging system.

As spectral data is often sequentially acquired, these systems tend to trade-off spectral and temporal resolution. The literature survey shows that spectral data are commonly acquired with a set of spectral bandpass filters in a mechanical filter wheel [4, 5] or a tunable filter, such as a liquid crystal tunable filter (LCTF) [4, 6, 7] or an acousto-optical tunable filter (AOTF) [8, 9]. Less frequently spectral data are acquired via a tunable Fabry–Pérot imaging spectrometer controlled by a micro-electro-mechanical system (MEMS) [10]. Spectral imaging is increasingly frequently also achieved via sequential illumination with narrowband light emitting diodes (LEDs) [11–13], or lasers [14, 15]. Whereas the sequential scanning of the spectral dimension limits the temporal resolutions of amplitude division imaging systems, sequential scanning can also be beneficial. When imaging temporally static objects the sequential acquisition allows for optimisation of the exposure time of each spectral band to maximise the achievable signal-to-noise ratio (SNR) [16]. The main advantage of amplitude division imaging systems is, however, that spectral data is not acquired at the expense of spatial resolution [17].

Spectral bandpass filters and tunable filters have long-standing commercial availability. Consequently, these types of filters have been used for several biomedical and clinical imaging applications [4–8]. Due to the limited amount of space in mechanical filter wheels, the use of spectral bandpass filters has, however, been limited to the acquisition of MSI data [18]. In contrast, both LCTFs and AOTFs have the advantage of providing tunable and programmable wavelength filtering for the acquisition of HSI data. A LCTF typically consists of a stack of Lyot filters, where each Lyot filter consists of a tunable birefringent liquid crystal plate placed between two linear polarisers [16]. The liquid crystal plates circularly polarise the light according to the thickness and birefringence of the liquid crystal plate and the wavelength of the incident light [16]. An electric field can be used to tune the birefringence of the liquid crystal plate to select the wavelength transmitted by the LCTF [16]. An AOTF consists of a crystal whose spectral transmission properties can be altered via the application of an acoustic wave; the frequency of the acoustic wave modifies the refractive index of the crystal such that it behaves as a transmission grating [16]. In an AOTF, diffractions occur over an extended volume such that light fulfilling the phase-matching criteria of the acoustically induced diffraction grating will be diffracted, whereas the remaining light propagates through the crystal undiffracted [16]. The diffracted light is transmitted through the AOTF while the remaining light is blocked by a beam stop [16]. The wavelength of the transmitted light can therefore be tuned by changing the frequency of the acoustic field [16]. Due to their wavelength flexibility and compactness, LCTFs and AOTFs are commonly used for HSI microscopy [17]. Both types of filters do however require linearly polarised light, which results in a 50% decrease in optical throughput when imaging unpolarised scenes [17].

LCTFs and AOTFs have been used both on the detection [6–8] and illumination side [9] of spectral imaging systems. With the increased availability and decreased cost of LEDs, narrowband LEDs are increasingly also being used to provide spectral illumination [11–13]. With further technological developments, spectral imaging via narrowband LED illumination does indeed have potential to become a cheap and flexible MSI method [11]. The biomedical applications [12, 13] of narrowband

LEDs are however typically somewhat limited, as temporal resolution tends to be sacrificed for spectral data acquisition.

Tunable Fabry–Pérot imaging spectrometers controlled via MEMS have the potential to be mass-produced and may therefore provide a cost effective approach to spectral imaging [10]. The ability to miniaturise this sensor—even allowing for package level integration into standard smart phones—enables the production of highly portable, compact and robust spectral imaging sensors [10]. As far as the author is aware, tunable Fabry–Pérot imaging spectrometers have however not yet been used in biomedical imaging applications. Future biomedical applications of tunable Fabry–Pérot imaging spectrometers may be limited as temporal resolution is sacrificed to acquire spectral data.

In contrast, MSI via lenslet arrays does not sacrifice temporal resolution for spectral data; a light field architecture—where a lenslet array is placed in front of the spectral filters—is instead used to spectrally divide the image [19]. Lenslet arrays can, for example, be used to replicate the image onto different spectral filters placed, either in front of [19], or behind [20, 21], the lenslet array. Whereas this provides for a compact MSI system with high temporal resolution, spatial resolution tends to be sacrificed since all spectral bands are generally imaged onto the same focal plane array (FPA). After the capability to directly deposit spectral filters on the pixels of the FPA was developed, MSI with lenslet arrays has become a less active area of research. Direct deposition of spectral filters on chip allows for a more compact and cost effective approach to MSI.

Aperture Division Spectral Imaging Systems

Aperture division spectral imaging systems acquire spectral data either by spatial division, or spectral sub-sampling, of the imaging system's aperture.

Although not traditionally classified as aperture division sensors, we assign HSI systems, which rely on relative motion between the sample and the sensor, to this category. The aperture of these sensors are divided, such that light from a point (x_i, y_i) or a line (x_i, y) is acquired at each spatial position of the sample. The light is then dispersed—using either a prism or diffraction grating—to acquire the spectrum of a point (x_i, y_i, λ; commonly referred to as a whiskbroom sensor) or the spectra of a line (x_i, y, λ; commonly referred to as a pushbroom sensor) [18]. The full HSI data cube is thus gradually constructed by spatial point or line scanning of the sample. The spatial and temporal resolution is typically dependent on the step size between spectral acquisitions. Temporal and spatial resolution is therefore traded with the spectral resolution.

Point and linescan sensors are well suited for applications with inherent movement between the sample and the sensor, for example conveyor belt inspection and remote sensing [18]. For applications without inherent movement between the sample and the sensor the need for spatial scanning is a serious limitation. Spatial scanning prevents video rate imaging, introduces motion artefacts, and the translation stages needed to move the sample or the camera make the systems bulky. There are, however, many examples of the application of point and linescan HSI systems for biomedical imaging, for example; tissue characterisation [22], diagnosis of gastric cancer from biopsies [23], and for confocal microscopy of membrane proteins [24]. Whereas long-term commercial availability partly explains the many biomedical applications of spatial scanning HSI systems, scanning systems can also acquire spectral image data cubes with high spatial and spectral resolution. The combined high spatial and spectral resolution make point and linescan sensors well suited for exploratory studies of the reflectance and/or fluorescence spectra of cells and tissues.

The aperture may also be spatially sub-sampled with a fibre bundle, where the back-end is reformatted into a linear array from which the output light is spectrally dispersed across a FPA [3]. This allows real-time acquisition of spectral and spatial image data [3]. Although there is no direct trade-off between critical imaging parameters, the light throughput of these systems are often limited due to inefficient light coupling [3]. In practice, the spatial resolution is also limited by the spacing of the fibrelets in the fibre bundle [3]. This approach is however promising for applications which require optical fibres independently of the spectral data acquisition, such as endoscopic imaging [25].

Real-time HSI can also be achieved through image mapping spectrometry (IMS) (also known as integral field spectrometry). In IMS, the image is spatially sampled at the aperture using a lenslets array or a set of faceted mirrors [26]. The spatially sub-sampled image is then spectrally dispersed and re-imaged onto a FPA [26]. This yields a compact system capable of real-time reconstruction of HSI data [26], well suited to biomedical imaging applications. Indeed, the group who first developed IMS has used the technology in microscopy [27], endoscopy [26] and retinal imaging [28]. It should however be noted that IMS sacrifices spatial for spectral resolution, as the image is sparsely spatially sampled to allow space to re-image the spectral dimension of the data cube onto a FPA.

Snapshot HSI Fourier Transform (SHIFT) is a computational technique which also relies on a lenslet array to form a set of sub-images [29]. This technique has, however, yet to be applied to biomedical imaging. In SHIFT, the sub-sampled images are Fourier transformed by passing them through two Nomarski prisms placed in front of the detector [3]. The Nomarski prisms consist of two birefringent crystal wedges cemented together [3]. By rotating the Nomarski prisms in relation to the detector, each image is exposed to a separate optical path difference (OPD) allowing the spectral data cube to be extracted via Fourier transform along the OPD axis [3]. Whereas a compact and robust SHIFT system with good spectral resolution can be realised [3], post-processing and extensive calibrations are required to extract the spectral data [30]. Despite a slight trade-off between spectral and temporal resolution (as the Normaski prisms need to be rotated) spectral data can however be acquired in real-time, provided the use of sufficiently short integration times [30]. The need for short integration times and linearly polarised light (as the OPD is introduced by birefringence effects in the prisms) reduce the optical throughput of the system [29]. SHIFT may therefore not be suitable to biomedical applications involving low light imaging.

Real-time aperture division MSI can be achieved with SRDAs [31]. The deposition of spectral filters in a mosaic pattern, directly in front or on top of the FPA, allows the full spectral data cube to be extracted from a single image frame [31].

The spectral filters may be created via deposition of stack material, either on a glass substrate or directly on top of the FPA, using simple lithography techniques [32]. SRDAs therefore provide a compact and robust approach to spectral imaging [32]. Clinical instrumentation is frequently moved in and out of storage, during which it may suffer bumps and bangs. Compact and robust imaging systems are therefore needed for clinical applications. In addition to being compact and robust, a majority of SRDA designs also have potential for mass-production and the associated low cost of manufacturing [32]. This makes clinical implementation into low resource settings conceivable. Indeed, although SRDAs have only relatively recently become commercially available, their use in a biomedical applications are already being explored [33, 34]. For example, SRDAs from Imec have also been applied in intraoperative neurological [35] and retinal imaging [34].

The main limitation of SRDAs is the acquisition of spectral data at the expense of spatial resolution. The spectral filters are deposited directly in front of, or on top of, the image sensor thus limiting its spatial resolution. Additionally, the spatial sub-sampling of the FPA reduces the photon efficiency of the sensor on which the spectral filters are deposited. Light at a given wavelength will be most efficiently captured by its wavelength matched spectral filter; the overall photon efficiency of the sensor is therefore inversely proportional to the number of spectral bands deposited on the FPA. In the literature, there are on the other hand several examples of work aiming to increase the spatial and spectral resolution of SRDAs via optimised spectral filter band selection [2, 36], spatial interpolation [2, 31] and compressed sensing techniques [37]. For further details on spectral filter band selection and interpolation techniques, the author refers the reader to the review by Lapray et al. [31].

Optical Path/Field Division Spectral Imaging Systems

Optical path division spectral imaging systems acquire spectral data by splitting the optical path of the light according to its wavelength, or to perform interferometry.

Use of dichroic filters is one of the most traditional ways to spectrally separate light into different optical beam paths. This method is also frequently used for biomedical imaging applications [38–41]. While dichroic filters allow diffraction limited imaging at high temporal resolution, the cumulative losses introduced by the filters do unfortunately limit the number of spectral bands in which imaging can be performed with an acceptable SNR [3]. The use of multiple dichroic filters also makes the systems vibration sensitive and bulky [3]. More compact filter based systems can be achieved through the use of filter stacks, although these designs increase cumulative transmission losses further [3].

Multiplexed volume holographic gratings (VHGs) provide a relatively compact system for MSI data acquisition [42]. A multiplexed VHG is a set of periodic phase or absorption perturbations in a volume media which, due to its 3D structure, is able to simultaneously diffract several target wavelengths by pre-designed angles [42]. This allows for video rate imaging in distinct spectral bands [42]. Biomedical imaging with a VHG has been demonstrated by fluorescence MSI of a mouse injected with fluorescent quantum dots [42]. The acquired images did however suffer from low resolution [43] and also required a trade-off between spatial and spectral resolution since the spectral bands were visualised on the same FPA [42].

Image replicating imaging spectrometry (IRIS) uses a set of birefringent interferometers, consisting of a wave-plate and a Wollaston prism [29]. Polarised light propagates through the wave-plate at a 45° linear polarisation relative to the birefringence axis of the wave-plate [29]. The incident light is thus split into two polarisation states along the ordinary and extraordinary axis of the wave-plate [44]. The polarization states aligned with the ordinary and extraordinary axis therefore experience different wavelength dependent refractive indices when propagating through the waveplate. Due to the polarization and wavelength dependence of the refractive index, the two polarization states therefore emerge from the waveplate with different phase, determined by the propagation distance in the waveplate and the wavelength of the incident light. The Wollastone prism, which has a phase and polarization dependent spectral transmission function, subsequently splits the light into two orthogonally polarized and divergent rays [44]. The spectral resolution of IRIS can therefore be increased to 2^N spectral bands by placing N birefringent interferometers in series [3].

Both IRIS and VHGs operate based on spectrally dependent redirection of the light and therefore do not suffer the transmission losses normally associated with spectral filtering [44]. In theory, the spectral resolution of IRIS can be increased without any associated light losses [29]. In practice, the high cost, chromatic aberrations and restricted availability of Wollaston prisms limit the achievable spectral resolution [29, 44]. Since the interferometer requires linearly polarised light, a light loss of 50% is also introduced when imaging unpolarised scenes [29]. Additionally, the same FPA tends to be used to image all spectral bands, such that spectral resolution is traded for spatial resolution. Nevertheless, the group who developed IRIS has proceeded to successfully exploit the real-time data acquisition of the method, to achieve spectrally resolved dynamic retinal oxiometry [45].

Several spectral imaging methods work based on interferometry. Image Fourier Transform Spectrometry—where one arm of a Michelson interferometer is scanned to sequentially obtain the spectral data—is the most traditional approach to image interferometry [3]. Since spectral bands are sequentially acquired, video rate data acquisition is not possible [3]. Michelson interferometers are also bulky and vibration sensitive, since the two arms of the spectrometer do not share a common beam path [3].

Other interferometry methods have been suggested to overcome the limitations of the Michelson interferometer, one approach is the multispectral Sagnac interferometer. Multispectral Sagnac interferometry uses multi-order blazed gratings to introduce wavelength dependent OPDs for specific wavelengths in an otherwise shared optical beam path [46]. This allows vibration robust real-time MSI interferometry [46]. In comparison to other spectral imaging systems, multispectral Sagnac interferometers are however bulky, only capable of acquiring MSI data [29] and have limited light throughput as they rely on polarised light [29, 46]. All interferometry techniques are also slightly more data intensive than direct acquisition of the spectral bands, as the acquired data needs to be Fourier transformed [29]. Hence, whereas Sagnac interferometers are commonly used in confocal microscopes [17], interferometry based systems are unlikely to be applied in bed-side clinical applications due to the aforementioned limitations.

Computational Spectral Imaging Sensors

Computational approaches to HSI have been developed to provide snapshot HSI without a hard trade-off of critical imaging parameters. The computational approaches to HSI tend to combine mechanically robust hardware components (to encode the spectral or spatial information) with computationally intensive data post-processing [3].

One of the main examples of computational HSI is Computed Tomography Imaging Spectroscopy (CTIS). In CTIS, a kinoform dispersing element spectrally diffuses the incident light onto a 2D sensor [3]. Spectral information is then obtained from multiple projections of the image at different viewing angles [3]. CTIS has the benefit of acquiring video rate high optical throughput HSI data from a compact instrument [47]. This makes CTIS well suited to biomedical imaging, as shown by its application in ophthalmology [47]. Future clinical applications of CTIS may, however, be limited by the need for extensive calibration and data post-processing [47].

Computational HSI via compressed sensing techniques is currently an active research area [37]. Compressive HSI sensing techniques rely on sparse hardware encoding of the spatial and/or spectral data followed by computational reconstruction of the HSI data cube [3]. A representative compressed sensing technique is the Coded Aperture Snapshot Spectral Imager (CASSI) [37]. CASSI uses a coded aperture to encode the spatial data before spectral dispersion of the light with a prism [37]. Recently there have been several modifications to CASSI in order to improve its spectral and spatial resolution, for example the development of the snapshot coloured compressive spectral imager (using a SRDA in the set-up), and dual coded compressive spectral imaging which separately encodes spatial and spectral data [37]. While a full HSI data cube can be acquired after data post-processing, a small image area and short data acquisition times is required for video rate imaging [37]. Compressive sensing techniques are computationally intensive and, although the algorithms for HSI reconstruction are constantly improved, meeting the associated hardware requirements have also proved challenging [3].

2.1.2 *Spectrally Resolved Detector Arrays*

In comparison to other spectral imaging systems, SRDAs are highly portable, compact, robust, and have the potential for very low manufacturing costs. This makes SRDAs particularly well suited for biomedical imaging applications. Prior to the widespread use of SRDAs for biomedical and clinical imaging applications, a few outstanding challenges do however need to be overcome. This section gives an overview of the general benefits and limitations of SRDAs for use in biomedical and clinical imaging applications. Several different approaches to realising SRDAs are also introduced.

General Benefits and Limitations of SRDAs

The concept behind a SRDA is very similar to that of a traditional RGB camera; spectral filters are deposited in a mosaic pattern on a FPA to allow spectral imaging at video rate. Whereas traditional colour cameras achieve spectral separation of light via organic dye based absorptive filters, alternative filter designs are required for multispectral imaging [48]. For example, organic dyes have broad spectral responses which may not be straightforwardly tuned [48]. Several alternative approaches to realising SRDAs have been explored; plasmonic filters, which achieve spectral selectivity via wavelength dependent resonant interaction of light with nanohole arrays [48, 49], or silicon nanowires [50, 51] and other resonant structures, such as thin film interference filters [52–54] and Fabry–Pérot interferometers [20, 32, 55]. The spectral selectivity of these notch and bandpass filters is tuned via material and/or geometric design choices [48, 56]. By assembling multiple filters on top, or in front of, the FPA compact and robust spectral imaging devices devices can be constructed [32, 56, 57]. Consequently, MSI/HSI via SRDAs require only limited calibration and alignment since spectral imaging is achieved on sensor, rather than via high level lenses or gratings [56, 57]. Additionally, as the spectral filters may be deposited in a mosaic pattern on sensor, the need for either spatial or temporal scanning can be sidestepped. This allow SRDAs to acquire MSI data at video rate [32]. The miniaturisability, portability, video rate spectral data acquisition and potential for low cost manufacturing make SRDAs highly suitable for clinical imaging applications.

Common to all mosaic type SRDAs is an inherent trade-off between spatial and spectral resolution. This trade-off further translates to a balance between spatial resolution and imaging SNR [31]. When the number of filters in a spectral macropixel increases, it also limits the opportunities to increase the SNR in a spectral band via standard methods, such as pixel binning [31]. Additionally, light at a given wavelength will only be efficiently captured by its wavelength matched spectral filter, such that the overall photon efficiency is inversely proportional to the number of spectral bands. The trade-off between spatial and spectral resolution is thus made in the sensor design process.

Ideally, the design of the mosaic filter pattern on the SRDA should be optimised to the specific imaging application [31]. The design of the mosaic pattern should optimally take into account; the distinguishing spectral features of the target and the illuminant, the QE of the underlying image sensor (i.e. the ratio of photogenerated electrons to the incident number of photons), the demosaicking method and subsequent image analysis and processing steps [31]. However, since SRDAs can typically not be spectrally tuned after assembly, custom sensors are required [31]. Due to the difficulties and cost of custom sensor manufacturing custom sensors are rarely realisable [31]. This is particularly problematic for small scale exploratory research projects, which do not benefit from the economies of scale [31].

To overcome the inherent trade-off between spatial and spectral resolution, many commercial manufacturers of SRDAs also provide linescan sensors [52, 57, 58]. Here, the spectral filters are deposited row-wise across the underlying image sensor (often incorporating several pixel rows [58]). Whereas video rate spectral imaging may not be performed with linescan SRDAs, these sensors allow true HSI. Up to hundreds of contiguous spectral filters may be deposited across the sensor without compromising spatial resolution. Linescan SRDAs may therefore be better suited for initial evaluations of the imaging performance of the filter technology, as high spatial and spectral resolution can be simultaneously maintained.

Different Approaches to Realising SRDAs

At present, MSI via plasmonic filters is the least mature SRDA technology [59]. Plasmonic filters relies on the interaction of light with plasmons i.e. the collective oscillations of free electrons in matter [59]. Plasmonic filters can be realised by controlling the plasmon resonances via nano-structural engineering of metal surfaces. Since the plasmonic response governs the spectral transmission/reflectance of the filter, the spectral selectivity of the filter can therefore be tuned by altering the geometries and dimensions of the nano-engineered material to support specific plasmon resonances [59].

There is currently a strong interest in developing complementary-metal-oxide-semiconductor (CMOS) compatible plasmonic filters, however, at present the most pioneering work has utilised non-CMOS-compatible noble metals (silver and gold) [59]. Aluminium is now, however, increasingly explored for the manufacturing of plasmonic filters, and is a promising candidate for large scale industrial production of plasmonic filters [59]. Currently, realisable plasmonic filters do, however, have low transmission efficiencies and spectral selectivity in the visible and NIR [60]. Nevertheless, with further developments in this rapidly emerging field, plasmonic filters show promise [61]. The nanostructures of plasmonic filters can for example be printed onto surfaces and show potential for very low cost manufacturing [61]. The spectral response of plasmonic filters are also independent of the AOI of the light; accessory optics may therefore be changed without affecting the spectral response of the sensor [51, 61].

Thin film interference filters, or realisations of Fabry–Pérot interferometers, are the most established methods for manufacturing of SRDAs [55]. Both filter types can be produced with CMOS compatible materials through standard lithography techniques, making high-volume commercial manufacturing feasible [32, 52, 54]. Although the exact design and materials used in commercial SRDAs are proprietary, the general designs are known.

Thin film interference filters typically consist of a combination of dielectric layers with alternating high and low refractive indices [62, 63]. The dielectric materials are combined to produce phase differences yielding constructive interference of the light within the passband of the spectral filter, whereas the transmission drops abruptly outside the transmission band [62, 63]. As shown by commercial vendors, such as Delta Optics [57] and Pixelteq [54], thin film interference filters with top-hat spectral transmission and high out-of-band light rejection can be produced.

Fabry–Pérot interferometers form high transmission [32] and narrow bandpass filters [62], by separating two reflecting surfaces by a spacer layer [62]. Light in resonance within the spacer layer will constructively interfere in the cavity, according to the following equation;

$$\frac{2\pi}{\lambda_0} \times nd\cos(\theta) - \frac{\phi_a + \phi_b}{2} = m\pi, \tag{2.1}$$

where m is the cavity order, ϕ_a and ϕ_b the phase shift after reflecting off the mirrors of the cavity, $nd\cos(\theta)$ the apparent optical thickness of the spacer layer, n the refractive index, d the spacer thickness and θ the propagation angle of light in the spacer layer [60]. Thus the thickness of the spacer layer controls the phase relationship which determines the peak wavelength of the transmitted light (λ_0) [60]. The FWHM and transmission efficiency of the spectral filters are governed by the reflectivity of the cavity mirrors [64].

It should be noted that the resonant structure of Fabry–Pérot interferometers leads to secondary transmissions at integer multiples of λ_0, causing significant side-band transmission [62]. Although this out-of-band transmission may be somewhat suppressed via the design of the cavity mirrors, edge pass spectral filters are typically used to block out-of-band transmissions [62]. To make the Fabry–Pérot interferometers CMOS compatible and manufacturable via standard lithography techniques, Bragg stack cavity mirrors are typically used [62, 65]. These mirrors are formed by alternating layers of dielectric with high and low refractive indices [62, 65].

2.1.3 The Imec and SILIOS Sensors

In this thesis, resonant commercial SRDAs incorporating Fabry–Perot interferometers have been studied. SRDAs based on Fabry–Pérot interferometers are arguably the most mature SRDA technology for spectral imaging [55]. The maturity of this technology makes it most promising for rapid integration of SRDAs into biomedical and clinical imaging applications. Although Fabry–Pérot filters have relative high out-of-band light leakage [62], they also allow spectral imaging with high transmission efficiencies [32], which is of particular importance to low light applications such as fluorescence imaging. The combination of the relatively narrow spectral filter response [62] with some spectral response tunability according to the AOI of the light [62] also allows for some imaging flexibility [58].

The Belgian company and research institute Imec is one of the commercial leaders in producing SRDAs. Imec is able to monolithically integrate Fabry–Pérot spectral filters at a pixel level onto standard CMOS sensors. In addition to the manufacturing and cost benefits of monolithic filter integration [32], wafer-level filter deposition also minimises spectral band cross talk and stray light generated by unwanted reflections of discrete components and substrate layers [32]. For the eventual use of SRDAs in the clinic, monolithic integration of the spectral filters appears suitable. The rest of this thesis therefore mainly focuses on the calibration and imaging evaluation of three Imec sensors; the NIR linescan sensor (Fig. 2.1a) and the visible and the NIR snapshot sensor (Fig. 2.1b).

Imec's linescan sensor (CMV2K-LS600-975-2.4.8.4; Imec) was, as far as the author is aware, the first commercially available SRDA with monolithically integrated resonant spectral filters. The linescan sensor consists of a Bragg stack Fabry–Pérot wedge filter deposited in a ladder structure on a CMOS sensor (CMV2000; CMOSIS), packaged in an Adimec camera system (Q-2A340m/CL; Adimec).

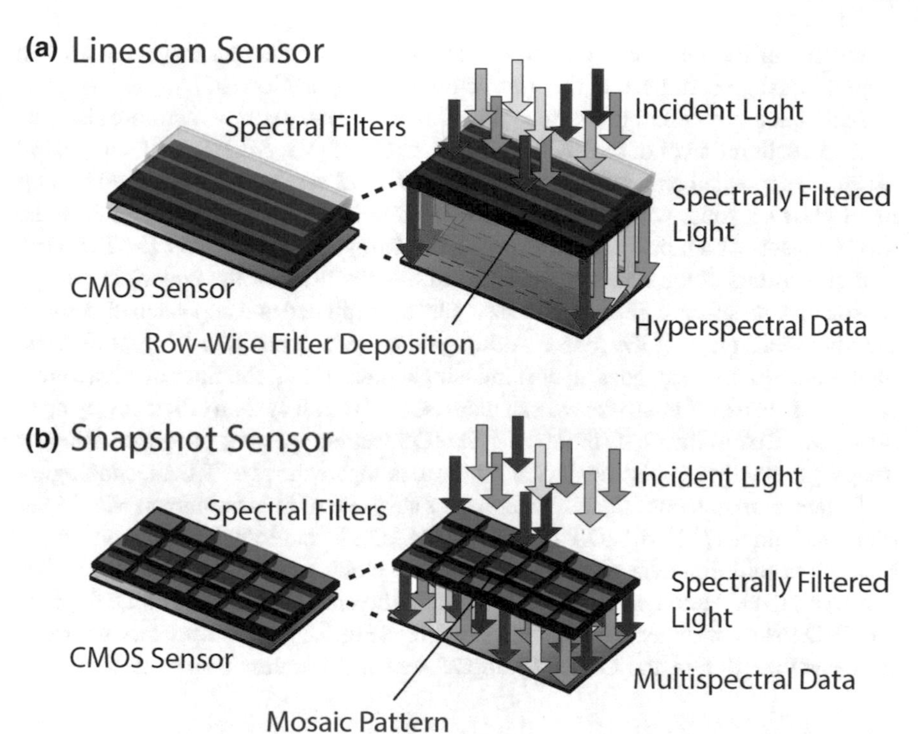

Fig. 2.1 Simplified schematic of a linescan hyperspectral and a snapshot multispectral SRDA. Spectral imaging of broadband incident light is achieved via spectral filters deposited on top of an image sensor. **a** For the linescan SRDA the spectral filters are deposited row-wise across the sensor and hyperspectral data acquired via spatial scanning of the sample. **b** The snapshot SRDA enables acquisition of multispectral data at video rate via deposition of the spectral filters on a pixel level in a mosaic pattern

The spectral filters have a nominal spectral range of 600–1000 nm, deposited row-wise across the sensor, such that each spectral filter band typically encompass eight pixel rows. With developments of the filter deposition technique, snapshot sensors based on the same technology became available during the process of this thesis. Two MSI cameras (visible, 460–630 nm: CMV2K-SSM4x4-9.2.10.3 and NIR, 600–1000 nm: SSM5x5 5.4.20.8; Imec) with 16 visible and 25 NIR spectral bands respectively are used in this thesis. These cameras were fabricated by deposition of filters in a 4×4 or 5×5 grid pattern on a pixel level directly onto CMOS sensors (CMV2000; CMOSIS) packaged in USB3 camera packages (MQ022HG; Ximea).

Although the monolithic filter integration pursued by Imec has several advantages, wafer-level filter deposition makes custom sensor design costly. Whereas one wafer may be used to manufacture thousands of image sensors in a cost effective manner, small scale manufacturing is expensive [32, 66]. The spectral filter deposition is one of the last lithography processing steps and the wafer therefore has high value compared to a clean piece of glass [66]. Depositing the spectral filters on clean glass may therefore be more suitable to small scale sensor production [66]. Glass deposition of the filters can also allow the end-user to select an image sensor and camera package optimised to the specific imaging application [67].

In addition to calibrating the monolithic Imec sensors, we have therefore also performed a calibration of the SILIOS MSI camera (CMS-V1; SILIOS), which spectral filters are deposited on a glass substrate. The SILIOS camera is constructed by gluing a glass substrate with spectral filters deposited in a 3 × 3 grid pattern onto the image sensors of a monochrome camera (NIR Ruby sensor, UI1242LE-NIR; IDS). The dimensions of the filters are matched to the pixel size of the underlying image sensor, and the spectral filters on the glass substrate aligned and deposited directly on the micro-lens array of the sensor. Although the direct deposition of spectral filters on the micro-lens array goes against industry standard [55], the filter manufacturers claim that the use of micro-lenses can increase the overall system efficiency by up to 50% [68]. Due to time constraints, the SILIOS camera was not integrated into any imaging evaluations, so is only included here to compare the two SRDA technologies.

Whereas Imec monolithically integrates Fabry–Pérot interferometers via lithography techniques [32], SILIOS' ©COLORSHADE technology "is based on a combination of thin film deposition and micro/nano etching process onto a fused silica substrate" [31]. Although the typical quantum efficiency (QE) curves of the filters of the SRDAs are made available by the suppliers (Fig. 2.2), the proprietary nature of the exact filter design and limited additional sensor calibration data make it difficult

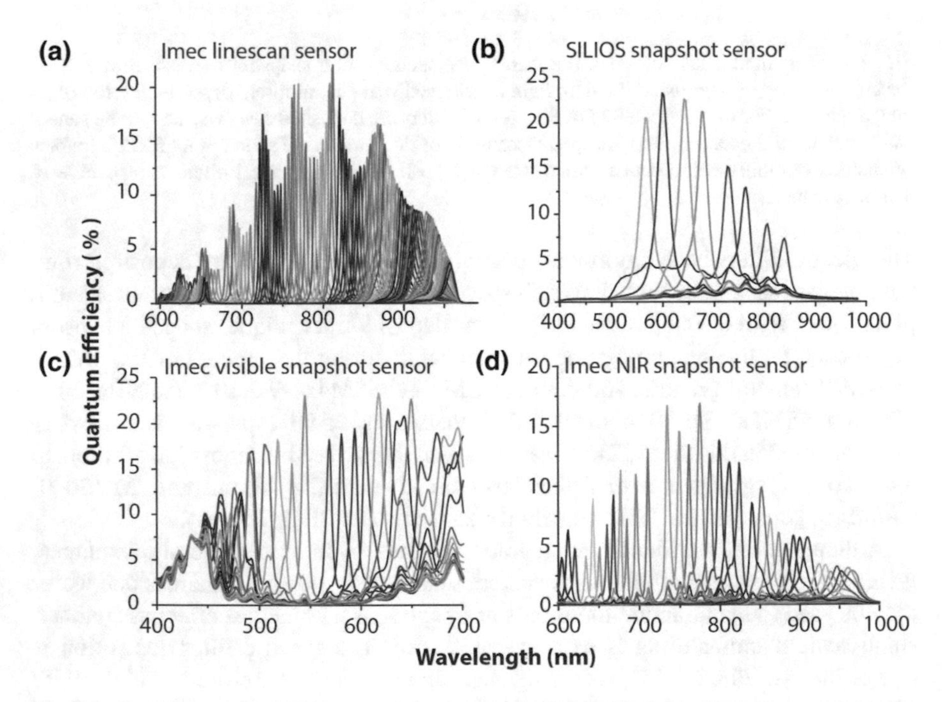

Fig. 2.2 Supplier provided QE curves of the sensors studied in this thesis. Each line represents the combined QE curve of the spectral filter and the underlying image sensor. Typical QE curves for the **a** Imec linescan sensor, **b** the SILIOS snapshot sensor, and **c** the Imec visible and **d** NIR snapshot sensor

to comment on the performance differences of the sensors without direct experimental comparison. We therefore performed in-house characterisations of the MSI cameras to enable direct comparison and independent evaluation of their spectral imaging performance. Standard spectral QE calibrations of the Imec linescan and SILIOS snapshot sensor were performed in-house. The two Imec snapshot cameras were not calibrated in-house, since the calibration data from the Imec linescan sensor convinced us that the QE calibration curves provided by the supplier were reliable. Additional to standard QE calibrations, the angular sensitivity of the spectral response of all four sensors was characterised in-house.

2.1.4 Standard Camera Calibration Methods

The QE of an image sensor is defined as the ratio of the photogenerated electrons to the number of photons incident on the sensor. This may be expressed with experimental parameters as;

$$QE(\lambda) = \frac{photogenerated\, electrons}{photons} = \frac{K_e \times DN(\lambda) \times hc}{I(\lambda) \times t_{int} \times A_{pixel} \times \lambda}, \tag{2.2}$$

where $DN(\lambda)$ is the digital number response of a pixel as a function of λ, K_e the conversion efficiency of the overall camera package, $I(\lambda)$ the illumination intensity incident on the pixel, t_{int} the integration time, and A_{pixel} the area of the pixel [69]. To agree with the European Machine Vision Association (EMVA) standard [70] (which is the accepted standard for camera calibrations), the camera should be calibrated with homogeneous pseudo-collimated light (F/# larger than 8) and the incident light intensity measured with a calibrated photo-detector. Throughout the calibration, all camera parameters should also remain constant [70]. The EMVA standard further assumes that; the camera response is linear (i.e. that the DN increases linearly to the number of photons received), noise sources are invariant with space and time, that only the dark current is temperature dependent, and only the total QE dependent on wavelength.

The number of photons can be straightforwardly determined via a measurement of the incident light intensity. The number of photogenerated electrons do however need to be extracted from the relative digital number (DN) of the sensor signal. The commercial SRDAs used in this thesis are based on CMOS sensors. A CMOS sensors consists of an array of identical pixels, each with a photodiode and an addressing transistor, of which each pixel row and column can be addressed via scan registers [71]. A photodiode consists of an electron rich (acceptor) and electron poor (donor) semiconductor material separated by an insulating (depletion) layer [71]. The charge difference between the acceptor and donor layer establishes a gradual voltage difference in the depletion region, which modifies the high (conductance) and low (valency) electron energy levels in the depletion layer. Via the photoelectric effect, a photon striking the depletion region of the diode can cause an electron to transition from the valency to the conductance band, such that a photocurrent is generated [71]. In a CMOS sensor, the photodiode is reverse biased and the generation of a photocurrent will decrease the voltage across the photodiode. At the end of the exposure time, the pixel is addressed and the decrease in the reverse voltage used to determine the number of photons which impinged upon the diode during the exposure time [71]. The voltage signal is then amplified and digitised by an analogue-to-digital converter (ADC) (either before or after pixel read-out) to produce the DN [70, 71]. The whole process may be assumed to be linear and treated as a black box defined by the overall conversion gain of the camera (K) expressed in DN/e^{-} [70]. The conversion gain of the camera is the inverse of the previously introduced conversion efficiency (K_e).

The conversion efficiency and gain of the sensor can be experimentally determined by taking advantage of its noise behaviour. The noise characteristics of the photo-generated electrons are fundamentally connected to the spatial manner in which the photons arrive at the sensor [69]. As the arrival of the photons is described by Bose-Einstein statistics, the noise of the photogenerated electrons is Poisson distributed; this is referred to as photon shot noise [69]. The variance of photogenerated electrons (σ_e^2) is therefore equal to the mean number of accumulated electrons (μ_e) [70];

$$\sigma_e^2 = \mu_e. \tag{2.3}$$

Following the previous assumption of signal linearity, the noise sources may be added and expressed in DNs according to;

$$\sigma_{signal}^2 = K^2(\sigma_{dark}^2 + \sigma_e^2) + \sigma_q^2, \tag{2.4}$$

to obtain the total signal variance (σ_{signal}^2). The photon independent noise ($K^2\sigma_{dark}^2$) and the quantisation noise (σ_q^2) can here be collectively referred to as readout noise (σ_R^2) [70];

$$\sigma_R^2 = K^2\sigma_{dark}^2 + \sigma_q^2. \tag{2.5}$$

Rewriting Eq. 2.4 we thus obtain;

$$\sigma_{signal}^2 = K(\mu_{signal} - \mu_{dark}) + \sigma_R^2, \tag{2.6}$$

where the raw camera signal has been dark-subtracted to isolate the photogenerated signals [70].

The linear dependence of the signal mean and its variance can be used to extract the conversion gain [70]. Data may be acquired via multiple acquisition methods which vary the number of photons incident on the sensor. This can be achieved, either by varying the intensity of the sensor illumination while keeping the camera integration time constant, or by keeping the illumination constant while varying the integration time [70]. If data is acquired by varying the integration time, the variance

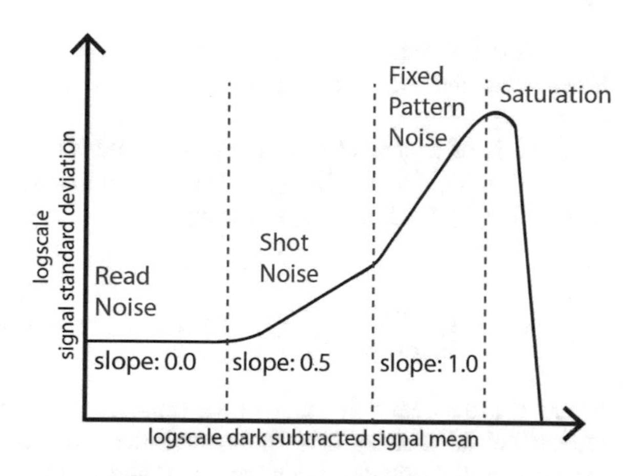

Fig. 2.3 An illustrative PTC showing the main noise regimes of a camera: read-out noise at low signal levels, followed by the shot-noise and fixed pattern noise region before camera saturation. The Figure has been modified from Bohndiek et al. [72]

of equivalent dark images does however need to be subtracted from the signal variance to compensate for integration time dependent variations in the read-out noise [70].

While the conversion gain may be determined by a mean-variance plot, the full noise behaviour of the sensor can also be studied via a photon transfer curve (PTC) [72, 73]. PTCs allow probing of the full noise characteristic of the sensor by plotting the signal standard deviation (σ_{signal}) against the dark subtracted mean signal ($\mu_{signal} - \mu_{dark}$) on a logarithmic scale (Fig. 2.3) [73]. At low signal levels, the standard deviation is dominated by the read-out noise; the readout noise of the camera can therefore be extracted from the y-intercept of the PTC [73]. Based on Eq. 2.6, the PTC can also be used to extract the conversion gain. The PTC is expected to have a slope of 0.5 in the shot noise dominated region, where the read-out noise may be considered negligible. The conversion gain can therefore be calculated from the x-intercept of a linear fit extrapolated from data within the shot noise region [73]. This method requires the slope of the PTC to be 0.5 prior to extrapolation in order to avoid the introduction of substantial errors when calculating the conversion gain [73]. The mean-variance technique is therefore more suited to extracting the conversion gain [73].

Prior to saturation of the camera, the PTC also shows a noise region dominated by fixed pattern noise (FPN) [72]. FPN refers to stationary inhomogeneities in the camera response, originating from dark signal non-uniformities (DSNU) and photo response non-uniformities (PSNU) of the pixels [70]. Whereas the FPN has not been characterised in this work, it should be noted that DSNU are removed by dark subtracting the acquired image frames [72].

2.1.5 Calibration Methods for SRDAs

A SRDA cannot be fully characterised using standard camera calibration methods, since spectral filters are integrated, on top or in front of, the image sensor [64].

In addition to standard parameters, calibration methods for SRDAs should also take into account the peak transmission and shape of the response curves of the spectral filters [20].

The DN of a spectral pixel ($DN_i(\lambda)$) depends both on the transmission of its filter ($T_i(\lambda, \theta)$) and the QE($QE_i(\lambda)$) of the underlying image sensor. This can be summarised by the following equation;

$$DN_i(\lambda) = \frac{K\ t_{int}}{hc} \times \int_{\lambda_1}^{\lambda_2} QE_i(\lambda) \times T_i(\lambda, \theta) \times A_{pixel} \times \lambda \times I(\lambda)\ \mathrm{d}\lambda, \tag{2.7}$$

where K is the conversion gain of the sensor, t_{int} the integration time and $I(\lambda)$ the intensity of the incident light and λ and θ the wavelength and AOI, respectively, of the incident light (the equation has been modified from that presented by Agrawal et al. [64] for increased generality). The equation highlights the angular sensitivity of resonant interference filters. For effective use of SRDAs, the spectral transmission characteristics of the sensor need to be understood; this includes the angular sensitivity of the spectral filter transmission.

The AOIs of the incoming light may be controlled by the F/# of the camera's objective lens. The F/# is defined as the ratio of the lens' focal length (f) to its aperture diameter (D) [48]. The F/# controls the half aperture angle ($\theta_{1/2}$) of the light falling on to the sensor according to [48];

$$\theta_{1/2} = \arctan\left(\frac{1}{2F/\#}\right). \tag{2.8}$$

Hence, when decreasing the F/# of a camera objective lens the AOI increases. This leads to a wavelength shift and widening of the spectral filter response [48, 74]. The broadening of the spectral filter response reduces the spectral selectivity of the filter [74]. However, because the light collected by an objective lens fall proportionally to the inverse square of its F/#, imaging at lower F/#s is also associated with higher light throughput [32]. At low F/#s, the combination of an increased optical throughput and a widened filter response can therefore be used to increase either the sensitivity or speed of spectral imaging [32]. The wide spectral filter response at low F/#s do, however, also decrease the spectral resolution [32]. Therefore, the choice of the objective lens includes an inherent trade-off between the spectral resolution and sensitivity of the SRDA [32].

It has since long been recognised that the AOI will impact the spectral filters' transmission [20, 62], yet the majority of sensors are still only calibrated with collimated light. Rather than calibrating and/or correcting for the angular dependencies of the spectral transmission, many commercial vendors and research groups have instead aimed to suppress the effects of incident angle by only imaging with pseudo-collimated light [20, 74].

Hardware has been used to ensure imaging with pseudo-collimated light and/or correct for variations in AOI. Gupta et al. [20] have, for example, used baffles, field stops and telecentric optics to restrict the AOI, whereas Imec recommends that imaging should be performed at an F/# equal or higher than 2.8 [74]. Frey et al. [60] further recognise that the AOI of the light will also vary across the FPA, and experimentally demonstrate a method to compensate for the blue shift of the filter response at oblique incidence, by combining plasmonic and Fabry–Pérot filter design. Whereas Frey et al.'s technique allows for filters with a constant transmission across the FPA, the filters need to be designed for a particular AOI which limits their application flexibility.

Software corrections and simulations of the angular sensitivity of the spectral filters have also been performed. The angular dependence of the filter transmission is often studied via thin film interference filter simulations by modeling the spectral response characteristics of the filters for a set of chief ray angles (CRAs) [53, 60, 64]. Recently Imec has also presented a software method able to "refine the calibration method for different set-ups" [33]. Imec's software method is implemented by imaging a set of reference targets with the SRDA and perturbing a calibration matrix—constructed from filter responses measured with collimated light—by a refinement matrix. The refinement matrix is found by minimising the differences between the acquired spectra of the reference targets (multiplied by the calibration matrix) and the known spectra of the targets. Since the correction matrix is only experimentally validated at an F/# of 2.8, it is not evident that this method can correct for the angular sensitivity of the filters. The method should, however, compensate for normal manufacturing variations of Imec's spectral sensors [33].

We have been unable to find examples of full experimental studies of the effect of F/# on the spectral transmission characteristics of SRDAs. One attempt to experimentally quantify the angular sensitivity of the spectral response of a RGB plasmonic sensor was developed by Burgos et al. [48]. Here, measurements of a Macbeth colour chart at a set of F/#s were performed to evaluate the impact of F/# on the colour representation of images acquired with the RGB sensor. However, as colour comparisons are limited to RGB data, this experimental approach is not directly translatable to multispectral SRDAs. To characterise the AOI sensitivity of the Imec and SILIOS sensors, we therefore designed a monochromator-integrating sphere set-up to investigate the impact of F/# on the spectral response of SRDAs.

2.2 Experimental Methods

2.2.1 QE Calibration of the Imec Linescan Sensor

A standard camera calibration of the Imec linescan sensor was performed to determine the position of the spectral filters on the sensor, and the combined QE of the filter response and the underlying image sensor. The supplier report states that the spectral filters are deposited row-wise across the 2M pixel sensor (2048 pixel columns × 1088 pixel rows, CMV2000; CMOSIS), such that each spectral filter encompasses approximately eight pixel rows. The pixel rows can therefore be grouped into so called 'spectral bands'. The supplier calibration report further states that the sensor has 100 spectral bands with a nominal spectral response of 600–1000 nm and FWHMs ranging between 4–12 nm. Thus 800 of the pixel rows of the sensor are used to extract spectral information. As the positions of the spectral filter bands were not specified by the manufacturer, the pixel rows on the CMOS sensor were here grouped and assigned to spectral bands, according to the similarity of their spectral response. After the position of the spectral bands had been determined, the conversion efficiency, read-out noise and dark current of the camera were extracted to allow calculation of the QE(λ) of the spectral bands.

QE Measurements

A commercial spectral calibration set-up (Aspect Systems) at the Rutherford Appleton Laboratory (Didcot, UK) was made available for sensor calibration. The calibration set-up (Fig. 2.4) contained a monochromator (grating line density 1200 g/mm) coupled to a broadband light source. A bi-convex lens (LB1723-B; Thorlabs) was used to expand and collimate the light from the monochromator, and the bare sensor was placed orthogonally to the collimated light. The set-up was enclosed in a light-tight box to minimise stray light.

A wavelength sweep of 600–900 nm, in 3 nm increments, acquiring 50 frames at each wavelength was performed. Dark frames were collected every 100 nm for dark subtraction of the data. Measurements were stopped at 900 nm, since the monochromator did not provide a stable output above this wavelength, and for the imaging evaluations performed we did not require the range of 900–1000 nm (Chap. 3). Throughout the measurement, the camera integration time was kept constant at 20 ms, the frame period at 40 ms, the gain at 1, and 2 tap acquisition of 10 bit images (without image averaging) was performed in linear mode (dynamic range of 60 dB). Prior to measurements, the camera was left running for a 40 min time period to allow its temperature to stabilise. The temperature did, however, drift from 60–63 °C during the measurement sequence. In response to the temperature increase, a decremental

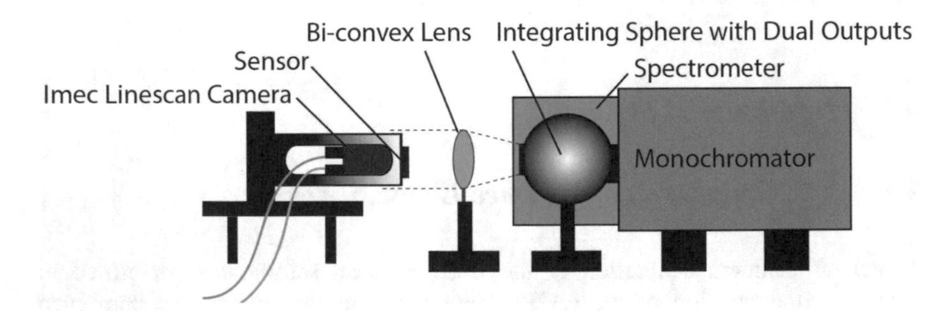

Fig. 2.4 Light from a commercial camera calibration system was used to extract the spectral response characteristics of the sensor. A bi-convex lens was used to expand and collimate the light from the monochromator, and the bare sensor placed orthogonally to the collimated light

wavelength sweep from 900–600 nm scan was additionally performed, during which the temperature increased further from 63–65 °C. Based on a qualitative comparison between the incremental and decremental wavelength scan, it was determined that the temperature drift had no noticeable impact on the dark subtracted data. We therefore only show data from the incremental scan. Following the measurements, the light intensity at the position of the sensor was calibrated with a supplier calibrated photodetector (N2772A MHz Differential Probe; Aglient), by sweeping the wavelength from 600–900 nm in 20 nm increments.

Data acquisition for all measurements with the Imec linescan camera were performed via the camera's graphical user interface (GUI) and a LabVIEW 2013 32-bit® acquisition code. It should be noted that the LabVIEW® image acquisition required linear stretching of the acquired images from 10 to 16 bit. The linear stretching affected the DN values in the subsequent analysis. All data analysis was performed in MATLAB 2013a, 2014a or 2015a®.

Identification of the Spectral Filter Bands

Identification of the positions of the spectral bands was achieved by identifying the pixel rows which responded similarly to collimated light. The spectral responses of the pixel rows were obtained by dividing the mean row pixel value of the average dark subtracted frames by the calculated number of photons incident on the pixel. The light intensity measured with the calibrated photodetector was extrapolated to intermediate wavelengths (to match the monochromatic light illuminating the sensor) by applying a polynomial fit to the acquired data. The light intensity was then converted to photons as in Eq. 2.2.

Conversion Gain, Read-Out Noise and Dark Current

To determine the conversion gain of the sensor, the bare sensor was separately illuminated with two broadband LED arrays with peak wavelength of 635 and 850 nm, respectively (Aspect Systems). During these measurements, the integration time of the camera was gradually increased until saturation of the relevant spectral bands was visually observed. 50 data frames followed by 50 dark frames were acquired at each integration time. Throughout the measurement the camera temperature increased from 56–57 °C.

Data were analysed according to standard analysis method detailed in Sect. 2.1.4. Mean-variance plots and PTCs were separately produced for each spectral band. To ensure similar illumination conditions, data were only extracted from spectral bands which reached saturation point when illuminated by the LEDs. Having determined the camera's conversion efficiency, the previously acquired response curves of the linescan sensor's spectral bands were converted to QE curves according to Eq. 2.2. The same conversion efficiency was used for all spectral bands, since it was assumed that the deposition of spectral filters did not impact the conversion efficiency of the underlying image sensor and camera package. This assumption was thought valid since the different spectral bands were observed to have similar conversion gain and read-out noise.

The linearity of the dark current was investigated by analysing the dark frames acquired during measurements with the 635 nm LED array. A linear fit was applied to the mean dark pixel values in each spectral band plotted against camera integration time.

2.2.2 *QE Calibration of the SILIOS Snapshot Sensor*

A standard QE camera calibration of the SILIOS snapshot sensor was performed. The SILIOS sensor (1280 × 1024 pixels, CMS-V-CMS16050007; SILIOS) has nine spectral filters bands deposited in a 3 × 3 grid pattern on a glass substrate; 8 narrow-band colour filters and a broadband neutral density (ND = 1) filter with a nominal spectral range of 550–830 nm [75]. The glass substrate is glued onto the CMOS sensor of a monochrome camera (NIR Ruby sensor, UI1242LE-NIR; IDS) such that the filter grid pattern is matched to the pixel pitch (5.3 μm) of the sensor [68].

In contrast to the Imec linescan sensor, an experimental investigation was not required to determine the position of the filter bands on the SILIOS snapshot sensor. The position of the spectral bands could instead be directly identified via visual inspection, owing to the regular grid pattern and the more spectrally distinct transmission characteristics of the spectral bands. SILIOS also shared their full calibration dataset, such that the conversion gain did not have to be measured in-house. A conversion gain of 8.4 e^-/DN and internal camera board gain of 0.5 (quoted by SILIOS) were thus used for data analysis.

QE Measurements

A QE calibration set-up was assembled in-house (Fig. 2.5). Light from a stabilised halogen light source (OSL2 with OSL2BIR bulb; Thorlabs) was coupled via a 0.48 numerical aperture (NA) and 1000 μm diameter optical multimode fibre to a monochromator (CM110 1/8m; Spectral Products) with a slit width of 0.15 mm and a grating (AG1200-00500-303; Spectral Products) with line density 1200 g/mm and blaze wavelength of 500 nm. Monochromatic light from an output fibre (M71L01;

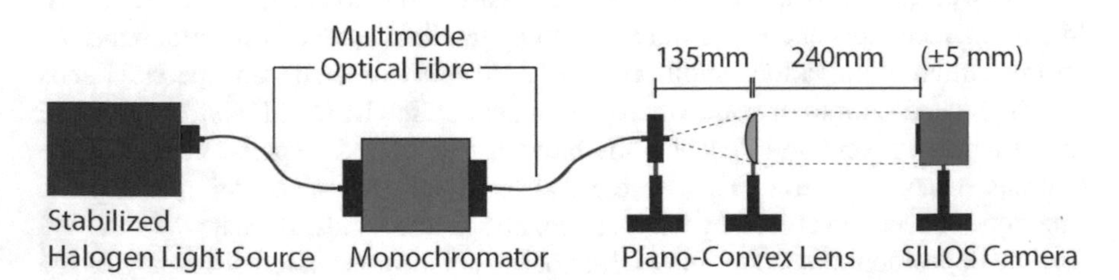

Fig. 2.5 Experimental set-up for QE Calibration of the SILIOS snapshot sensor. A spectral scan between 400–980 nm was performed by coupling light from a stabilised halogen light source to a monochromator via an optical fibre. Monochromatic light from the output optical fibre was collimated by a plano-convex lens before illuminating the sensor

Thorlabs) was collimated using a achromatic doublet lens (AC254-150-A; Thorlabs) before illuminating the sensor.

The quality of the collimation was qualitatively checked by allowing the beam to propagate in free space for approximately 5 m; the beam size did not notably change over this distance, indicating pseudo-collimated light. The temporal stability of the collimated light source was characterised by placing a supplier calibrated photodetector (1916-R; Newport) at the position of the camera. The wavelength of the monochromator was set to 670 nm and the light intensity recorded in 10 s intervals over a 30 min time period, during which a 3% light intensity variation was observed, representing the largest observed random error in the experimental parameters. This was considered sufficiently low to enable an accurate comparison between the supplier provided and in-house measured camera calibration curves.

A full spectral scan was subsequently performed with the calibrated photodetector placed in the position of the camera. The wavelength was scanned between 400–900 nm and the light intensity recorded every 10 nm.

The SILIOS sensor was placed in the beam path and a spectral scan performed between 400–980 nm in 3 nm increments, acquiring 20 frames at each wavelength. 80 dark frames were acquired prior and after the spectral scan to allow for dark subtraction of the data. The camera parameters were set in an initialisation file, using the camera's GUI (uEYE cockpit; IDS GmbH). The initialisation file was subsequently loaded into the automatic LabVIEW® acquisition code used to control the camera and the monochromator. Throughout the spectral scan, the camera exposure time was kept constant at 680 ms, with a master gain of 4 (corresponding to level 100 in the GUI) and 8 bit frames, without image averaging, were acquired. As the camera package did not include a temperature sensor, the camera's temperature could not be monitored; the camera was instead allowed to stabilise for 10 min prior to starting the data acquisition. The ambient temperature increased from 23–24 °C during the spectral scan.

The dark frames acquired prior and after the spectral scan were averaged and used to dark subtract the data. The spectral pixels were separated into their respective spectral bands, and the average DN(λ) of each band extracted. DN(λ) was converted to QE(λ) using Eq. 2.2. The camera gain function was included by dividing the expression in Eq. 2.2 by the total sensor gain (master gain $\times$ board level gain). The light intensity measured with the calibrated photodetector was extrapolated to obtain the light intensities at wavelengths intermediate to the experimentally acquired data by applying a polynomial fit.

2.2.3 F Number Calibrations of SRDAs

An experimental approach to study the effect of F/# on the spectral transmission characteristics of the SRDAs was developed. A monochromator-integrating sphere system (Fig. 2.6) was assembled to allow measurements of the sensors' spectral response curves at a set number of camera objective F/#s using a variable

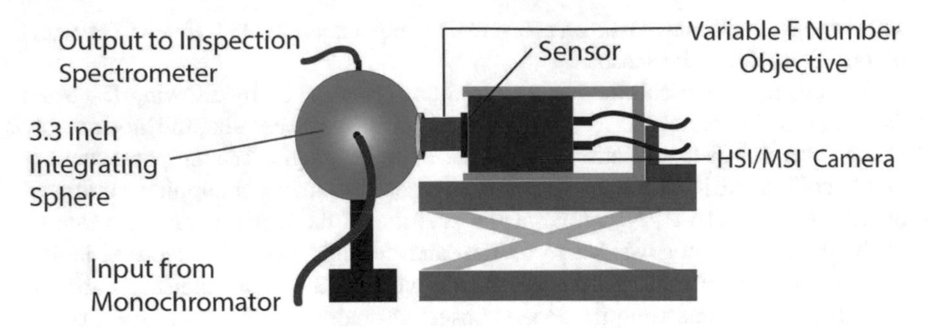

Fig. 2.6 A monochromator-integrating sphere system was designed and assembled to investigate the F/# dependence of the spectral transmission characteristics of the SRDAs

wide-field objective (35 mm VIS-NIR compact fixed focal length lens; Edmund Optics). A broadband halogen light source (OSL2 with OSL2B2/OSL2BIR bulb; Thorlabs) was coupled via a 0.48 NA 1000 μm diameter fibre (M71L01; Thorlabs) into a monochromator (CM110 1/8m; Spectral Products) with a slit width of 1.2 mm and a grating (AG1200-00500-303; Spectral Products) with line density 1200 g/mm and blaze wavelength of 500 nm. The monochromatic light was relayed via an optical fibre (M71L01; Thorlabs) to the integrating sphere (AS-02994-033; Labsphere). Isotropically diffuse light output from the integrating sphere illuminated the camera objective at the 1.5" output port of the integrating sphere. The focal length of the objective lens was kept at infinity throughout the measurements to avoid imaging the internal structure of the integrating sphere. A calibrated inspection spectrometer (AvaSpec-ULS2048-USB2-FCDC; Avantes) was coupled to the third output port of the integrating sphere to allow for concurrent measurements of the intensity spectra.

We defined the 'relative spectral response' recorded with the objective lens in place as;

$$SR = \frac{DN}{I_{max} \times \lambda \times t_{exp}}, \tag{2.9}$$

where SR is the relative spectral response of the camera directly extracted from experimental parameters, t_{int} (μs) the integration time, λ (nm) the peak wavelength of the spectral line from the monochromator, and I_{max} (μW cm^{-2}) the peak intensity recorded by the inspection spectrometer. Since the F/# calibrations do not conform to the EMVA standard (Sect. 2.1.4), we here use the SR rather than the QE. The SR is here directly expressed in terms of experimental parameters, yielding units of $DN\,\mu W^{-1}\,cm^2\,nm^{-1}\,\mu s^{-1}$.

F Number Calibration of the Imec Linescan Sensor

The relative spectral response of Imec linescan camera was measured with the objective lens and an additional 600 nm IR longpass filter (Imec) placed in front of the camera to replicate the optical path later used in the fHSI wide-field system (Chap. 3).

The F/# of the objective lens was set to 1.65 and a spectral scan was performed between 580–920 nm in 3 nm increments, acquiring 50 frames at each wavelength. 50 dark frames were acquired prior and after the spectral scan to allow for dark subtraction of the data. The camera integration time was set to 99.97 ms, the frame period to 100 ms, the gain to 1 and 4 tap acquisition of 10 bit images were performed. Prior to the measurements the camera's operating temperature was allowed to stabilise. Throughout the measurements the temperature of the camera drifted by a maximum of 2 °C. Keeping the camera parameters constant, the spectral scan was repeated for an objective F/# of 2.0, 2.8 and 4.0. The dark frames acquired prior and after the spectral scans were averaged and used to dark subtract the acquired data. Dark subtracted data were converted to relative spectral response curves according to Eq. 2.9.

F Number Calibrations of the Snapshot Sensors

Spectral scans with the variable camera objective lens F/#s of 1.65, 2.0, 2.8 and 4.0 (NIR longpass filter removed) were performed for the SILIOS snapshot sensor. The wavelength range of 380–920 nm was scanned in 3 nm increments, acquiring 20 frames at each wavelength. 8 bit frame acquisition, without image averaging and with a master gain of 4, was performed. 50 dark frames were acquired prior and after the scan. The exposure time was set to 300 ms for the spectral scans at F/#s 1.65 and 2.0, 500 ms for a F/# of 2.8 and 650 ms for a F/# of 4.0. The dark subtracted, averaged data were converted to relative spectral response curves according to Eq. 2.9.

Spectral scans and data analysis following the same experimental protocol were performed for the Imec visible and NIR snapshot sensors. The communication protocols of the Imec cameras were switched to USB3 via the Ximea GUI (XiCop; Ximea). 10 bit frame acquisition, without image averaging and a gain of 0.0 (set in *NIMax* prior to frame capture via the custom Labview® data acquisition code), was performed. A high power halogen bulb (OSL2B2; Thorlabs) was coupled to the monochromator to perform the spectral scans. For the Imec visible snapshot camera the exposure times were set to 100 ms for F/#s 1.65 and 2.0, 180 ms for F/# 2.8 and 320 ms for F/# 4.0. For the NIR Imec camera the exposure times were set to 950 ms for F/#s 1.65 and 2.0 and 999.5 ms for F/# 2.8 and 4.0. The acquired data were converted to relative spectral response curves according to the previously described method.

For the spectral scan of the imec NIR snapshot sensor at a F/# of 2.8, a significant baseline was however noted in the output spectra from the monochromator (recorded by the inspection spectrometer). The increased light leakage from the monochromator may be due to erroneous resetting of the monochromator grating between successive wavelength scans. Due to the increased baseline in the monochromator spectra, this dataset was not further analysed. Instead, measurements at three F/# scans were deemed sufficient to determine the spectral response's dependence on F/#.

2.3 Results

2.3.1 *QE Calibration of the Imec Linescan Sensor*

Identification of the Spectral Filter Bands

The pixel rows of the Imec linescan sensor were grouped and assigned to spectral bands based on their spectral response to collimated light. The active region of the linescan sensor could be approximately determined from the step-wise linear increase of the peak wavelength with row number (Fig. 2.7a). The similarities of the pixel rows' spectral responses were used to determine the position and number of pixel rows in each spectral band. The pixel rows were grouped into bands of seven, eight and nine rows and shifted row-wise up and down from row number 345 (Fig. 2.7b). To quantify the intra-group similarity, the standard deviation of the peak wavelength, the FWHM and the intra-group city block metric [76] of different pixel rows groupings were calculated. Based on the lowest obtained intra-group city block metric, each spectral band was found to contain eight pixel rows, positioned such that a new spectral band starts at row 345. This result is further supported by the behaviour of the intra-group standard deviation of the peak wavelengths and FWHMs of the spectral responses (data not shown). 72 spectral bands between pixel rows 329-905 with spectral response peak wavelengths between 600–900 nm were thus identified.

After having identified the spectral band positions, the QE curves of pixel rows within the same spectral band were qualitatively inspected. Pixel rows at the edges of the spectral bands were observed to have lower QE than the remainder of the pixel rows within the same spectral band (Fig. 2.7c). After further discussion with the manufacturer, this edge effect was attributed to a slight spatial mismatch between the filter edges and the pixel row edges. This was accounted for by excluding pixel rows at the edges of the spectral bands in subsequent data analysis and presentation.

Conversion Gain, Read-Out Noise and Dark Current

Based on the gradients of the mean-variance plots (Fig. 2.8a), the Imec linescan sensor was found to have a conversion gain of $6.0 \pm 0.3\ DN/e^-$, and therefore a conversion efficiency of $0.16 \pm 0.01\ e^-/DN$. A read-out noise of 95.00 ± 0.07 DN was further determined from the y-intercept of the PTCs (mean $\pm$ standard error across spectral bands; Fig. 2.8b).

The dark current increases linearly with integration time ($R^2 \geq 0.999$ for all bands; Fig. 2.8c). The dark current was measured as 5500 ± 2000 DN/s with an y-intercept of 2135 ± 6 DN. It should, however, be noted that the camera was not temperature controlled during these measurements. Since dark current is expected to have an exponential dependence on the camera operating temperature, a full dark current characterisation would therefore require additional measurements at multiple operating temperatures [70]. The dark current was here calibrated at a typical operating temperature. Although not a full dark current characterisation, the quoted value can be considered representative for standard camera operation.

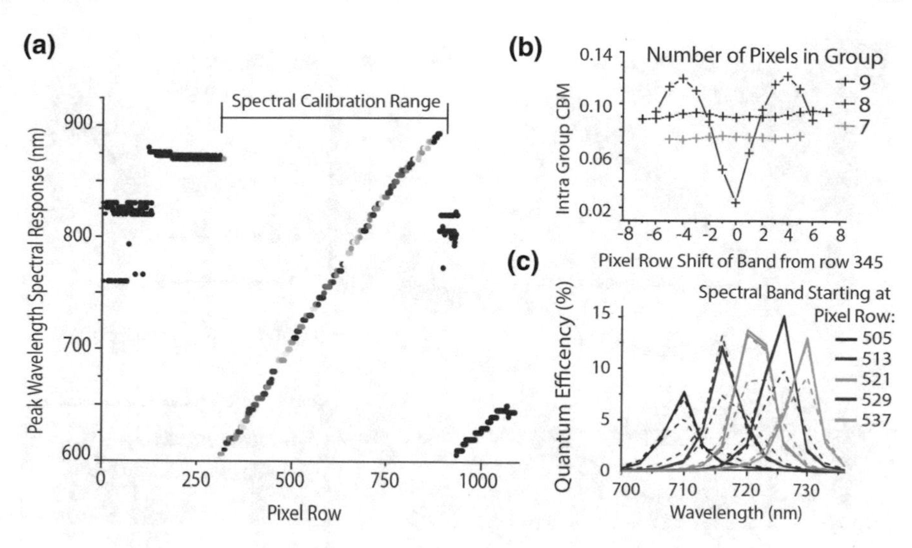

Fig. 2.7 **a** The active region of the Imec linescan sensor could be approximately determined from the step-wise liner increase of the peak wavelength response with pixel row number. The colour coding of the data points indicate the final grouping of the pixel rows into spectral bands. The black data points show pixel rows either outside the active region of the sensor, or outside the spectral calibration range. **b** The position and number of pixel rows in a spectral band were determined by the minimum intra-group city block metric (CBM) of the pixel rows arbitrarily grouped into bands. The position of the bands was shifted up and down from row 345. The lowest intra-group city block metric indicates that each spectral band contained 8 pixel rows, and that the bands should be positioned such that a new spectral band starts at row 345. **c** The spectral response curves of pixel rows assigned to spectral bands. Pixel rows at the edge of the spectral bands are indicated by a dashed line. These pixel rows were observed to have a lower QE than the remainder of the rows in the band

Camera parameters extracted from different spectral bands had similar values (Fig. 2.8). We can therefore assume that the deposition of spectral filters do not impact the conversion gain, read-out noise or dark current of the underlying CMOS sensor. A single value is therefore quoted for the conversion efficiency, read-out noise and dark current. These values are considered valid for all spectral bands.

QE Calibration Results

72 spectral bands were identified with peak wavelengths in the 600–900 nm spectral range. The peak QEs vary between 1.5–15.9%, and the FWHMs between 11–18 nm (Fig. 2.9). The root-mean-square-error (RMSE) of the experimentally measured bands, in comparison with supplier provided data, was found to be $1.4 \pm 4.0\%$ (mean $\pm$ standard error across spectral bands; experimental data interpolated to match supplier provided data), which agreed with the supplier provided data within the experimental error margin.

Whereas the low RMSE indicates that the spectral band responses measured in-house were aligned with the supplier provided QE data (Fig. 2.2a), qualitatively it was observed that the in-house measured QEs were systematically lower than the

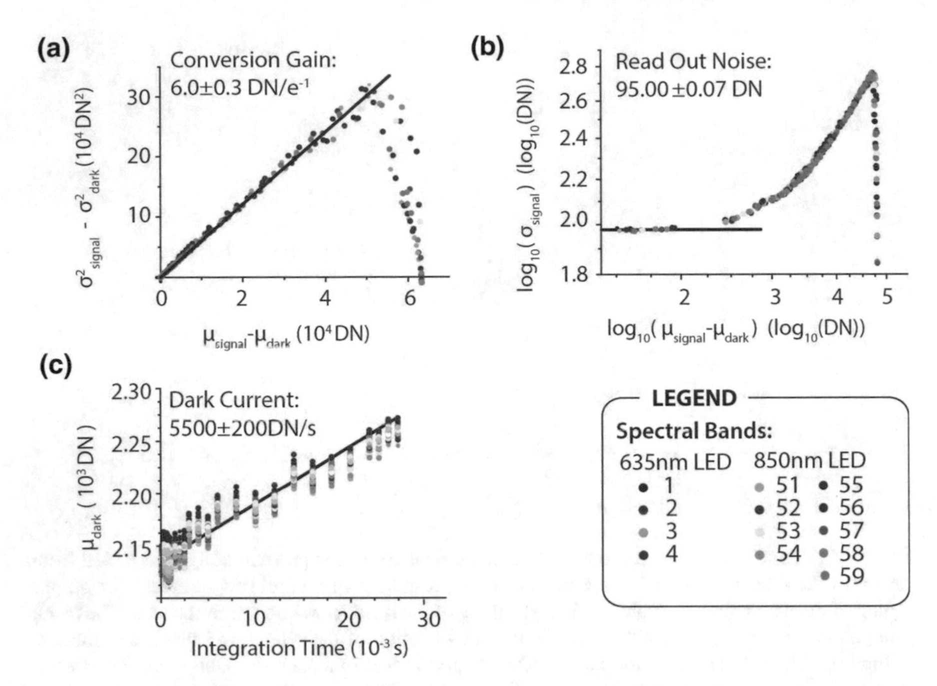

Fig. 2.8 Plots used to extract the conversion gain, read-out noise and dark current of the Imec linescan sensor. **a** The conversion gain was extracted from the gradient of the mean-variance plot, **b** the read-out noise from the y-intercept of the PTC, **c** and the dark current from the gradient of the mean dark DN value plotted against integration time. Data were only extracted from the spectral bands which were saturated during the experiment (indicated by the figure legend)

supplier provided data. The discrepancy between the absolute QE values may be due to manufacturing differences between sensors or differences in the calibration set-ups. For example, whereas the supplier calibration was performed in 1 nm wavelength increments, 3 nm wavelength increments were used in the in-house calibration. It is therefore possible that the response of certain spectral bands was not sampled frequently enough to access their peak wavelength response. Considering the low RMSE, we can however conclude that the in-house QE calibration (Fig. 2.9) yielded results aligned with supplier provided QE data (Fig. 2.2a).

2.3.2 QE Calibration of the SILIOS Snapshot Sensor

Periodic Noise Pattern in SILIOS Spectral Band Images

A periodic noise pattern was observed in spectral band images acquired with the SILIOS snapshot sensor when illuminated with monochromatic light. The sensor manufacturers attribute this periodic noise pattern to interference effects between the glass substrate and the CMOS sensor [68]. In contrast, the pattern was not present in

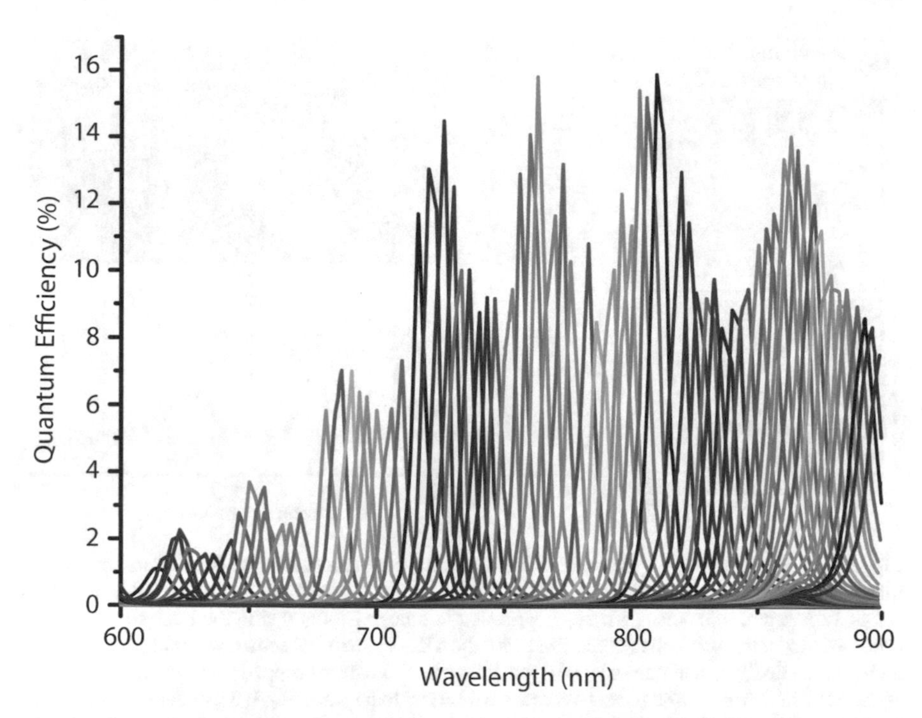

Fig. 2.9 Spectral band QE curves obtained from the experimental calibration of the Imec linescan sensor. 72 spectral bands were identified with peak wavelengths within the 600–900 nm spectral range. Each line shows the QE(λ) of one spectral band (the same colour is used for several bands). The in-house QE calibration yielded results aligned with supplier provided QE data (Fig. 2.2a)

spectral band images acquired with the Imec sensors, for which the spectral filters are monolithically integrated (Fig. 2.10a).

The periodicity of the observed pattern changed with wavelength (Fig. 2.10b), as would be expected from interference effects between the glass substrate and the sensor. Indeed, the pattern resembles Newton rings in transmission—with a bright central region surrounded by concentric interference fringes—which could typically appear due to light interactions in a cavity formed between a spherical and a flat reflecting surface [77]. According to these observation, it is plausible that a model of the interference could be used to directly correct the observed noise pattern. This would, however, require an additional calibration step to first fully characterize the expected wavelength dependence of the noise pattern. Therefore, whereas we note that the noise pattern may be problematic for future imaging applications, we here averaged the spectral band images to extract the average QE(λ) response of the sensor. Extracting the average spectral band response was thought appropriate, as the noise pattern appears as modulations on an otherwise flat intensity profile.

QE Calibration Results

Figure 2.11 shows the in-house measured QE curves of the SILIOS snapshot sensor. The peak QEs of the spectral bands was found to vary between 22–28% and the FWHM between 24–27 nm, showing good agreement with the supplier provided data

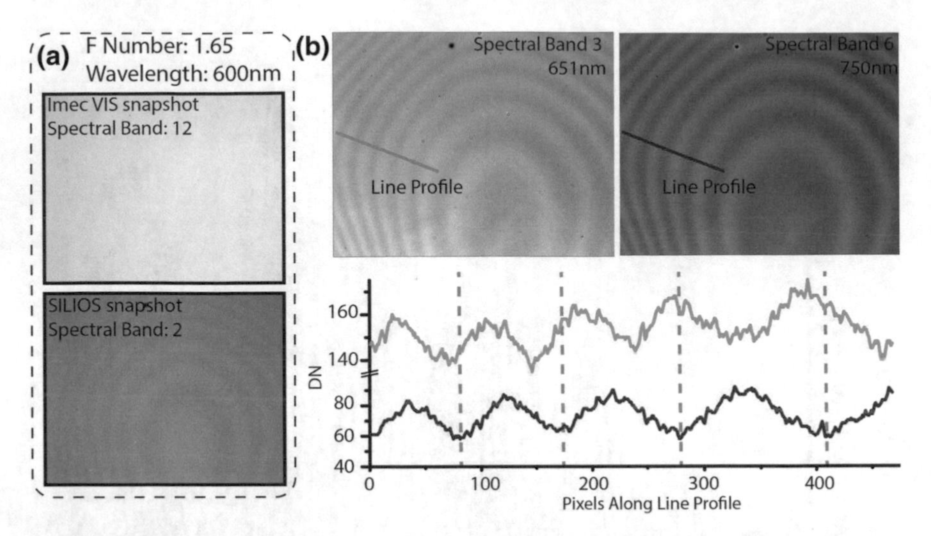

Fig. 2.10 **a** Representative spectral band images from the SILIOS and the Imec snapshot sensors illuminated with monochromatic light. The images were acquired with a camera objective lens F/# of 1.65. A periodic noise pattern—not present in images acquired with the Imec sensors—was observed in spectral band images acquired with the SILIOS snapshot sensor when illuminated with monochromatic light. The Imec spectral band image has here been cropped to match the form factor of the SILIOS sensor. **b** The noise pattern was observed to be wavelength dependent, as illustrated by line profiles extracted from spectral band images acquired with the SILIOS sensor illuminated with 651 and 750 nm monochromatic light, respectively

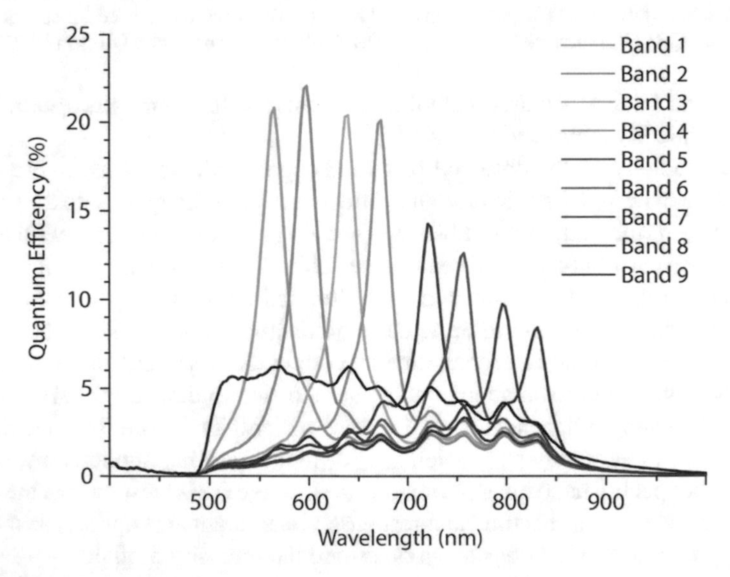

Fig. 2.11 Experimentally measured QE curves of the spectral bands of the SILIOS snapshot sensor. Each line shows the QE(λ) of one spectral band. The in-house QE calibration yielded results aligned with supplier provided QE data (Fig. 2.2b)

(Fig. 2.2b). The RMSE of the experimentally measured QE curves, when compared to supplier provided data, was 3.8 ± 0.7% (mean ± standard error across spectral bands; experimental data interpolated to match supplier provided data). This error slightly exceeds the experimental error margin. The relatively high RMSE values here appear to originate from disagreements between the out-of-band transmission of the supplier provided and in-house calibration data.

As previously stated, the conversion gain of the SILIOS snapshot sensor was provided by the supplier. The conversion gain, read-out noise and dark current were therefore not measured in-house.

2.3.3 F Number Calibrations of SRDAs

To study the effect of AOI on the spectral transmission characteristic of SRDAs, a monochromator-integrating sphere system was assembled, allowing measurements of the spectral band response at a set of camera objective lens F/#s. The relative spectral response (Eq. 2.9) was calculated and the following metrics extracted to quantitatively describe the changes in spectral band response with F/#; the peak wavelength of the response, the FWHM, the total area-under-curve (AUC), and the AUC within the FWHM. Datasets with strong spectral baselines were baseline subtracted prior to extracting quantitative metrics.

All data in this section are quoted as the mean ± the standard deviation across the spectral bands.

F Number Calibration of the Imec Linescan Sensor

The response curves of the spectral bands of the Imec linescan sensor (Fig. 2.12a–d) were observed to have strong F/# dependent spectral baselines (Fig. 2.12e). The spectral baseline was higher for lower F/#s, indicating a decreased out-of-band optical density of the spectral bands at oblique incidence angles. A decreased out-of-band optical density with oblique incidence angles is indeed expected as the imec linescan sensor uses Fabry–Pérot interferometers. Each spectral filter is therefore designed for a specific spectral range, when outside of the specified spectral range (as allowed by operating at oblique incidence angles) the reflectivity and finesse of the cavity mirrors typically decrease, which leads to higher out of band light leakage [77].

It is however difficult to comment on the wavelength dependence of the observed baseline. A relatively large monochromator slit width of 1.2 mm, yielding spectral lines with an average FWHM of 8 nm, was used during the spectral scan to maximise light throughput at the expense of spectral resolution. This yields a significant uncertainty in both λ and I_{max}, which may impact the baseline structure significantly. Any uncertainty in λ and I_{max} is however independent of F/#; the relative AUC of the baseline may therefore be compared across different datasets. As the F/# of the objective lens changed from 1.65 to 4.0, the AUC of the baseline decreased by 84%.

For quantitative comparison of the in-band relative spectral response (Fig. 2.13), the spectral response curves were baseline subtracted. As the F/# of the objective lens

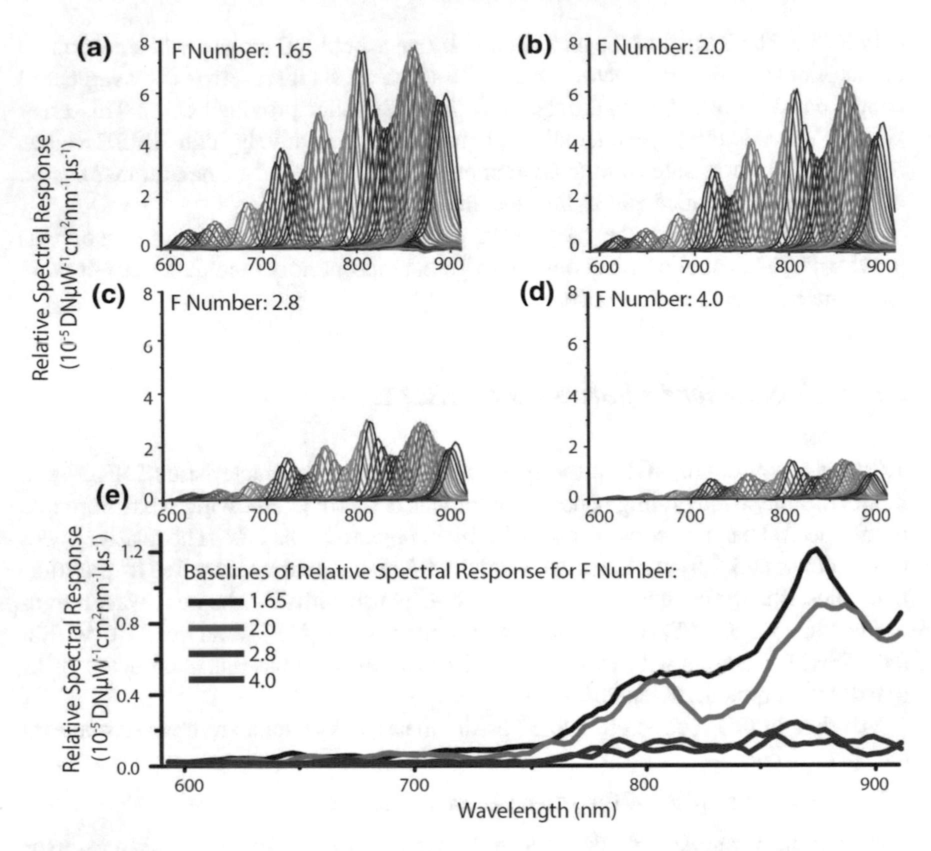

Fig. 2.12 Baseline subtracted relative spectral response curves of the Imec linescan sensor for an objective F/# of **a** 1.65, **b** 2.0, **c** 2.8 and **d** 4.0. **e** The F/# dependent baselines of the relative spectral response curves

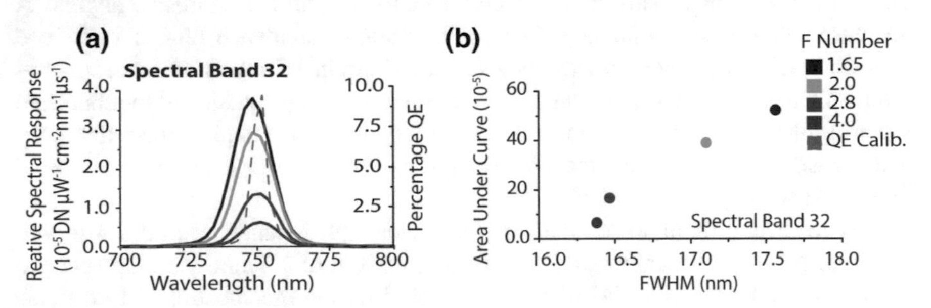

Fig. 2.13 **a** The F/# dependence of the spectral transmission characteristics of the Imec linescan sensor shown for a representative spectral band. As the F/# of the objective was decreased, the centre wavelength of the relative spectral response converged on that of the in-house measured QE (dashed grey line). **b** At lower F/#s the FWHM of the spectral response became broader as the AUC increased, indicating a higher optical throughout at lower F/#s

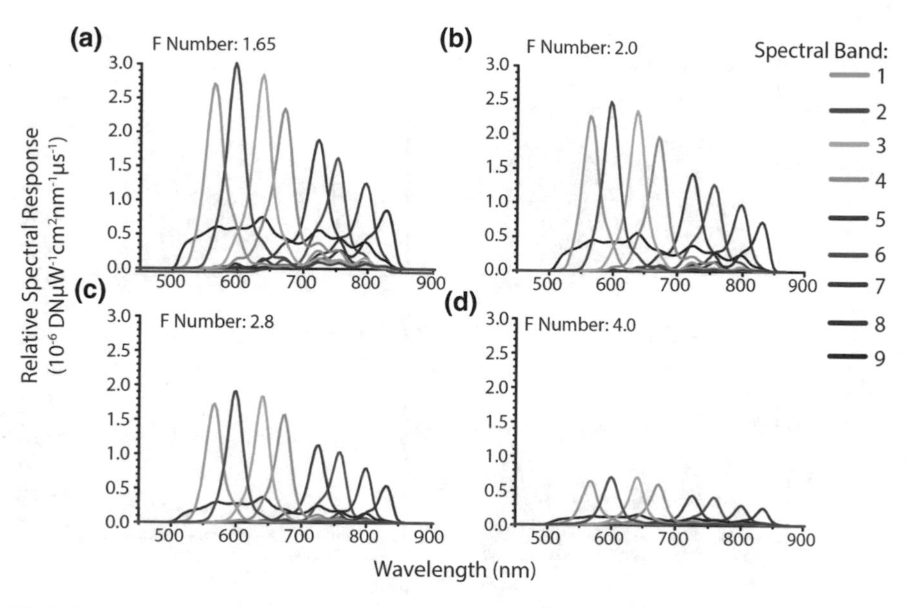

Fig. 2.14 The relative spectral response curves of the SILIOS snapshot sensor for an objective lens F/# of **a** 1.65, **b** 2.0, **c** 2.8 and **d** 4.0

changed from 1.65 to 4.0, the optical throughput of the baseline subtracted data—as characterised by the AUC—decreased by 84 ± 1%. Meanwhile the FWHM of the spectral response decreased by 8 ± 2% (Fig. 2.13). This highlights the F/# dependent trade-off between signal sensitivity and spectral resolution. As the F/# changed from 1.65 to 4.0, the corresponding shift in peak wavelength was 2.9 ± 1.5 nm. According to expectation, the centre wavelength converged with that measured using collimated light as the camera objective lens F/# was increased.

F Number Calibration of the Snapshot Sensors

The response of the spectral bands of the SILIOS (Fig. 2.14), Imec visible (Fig. 2.15) and Imec NIR (Fig. 2.16) snapshot sensors all show strong dependence on the camera objective lens F/#. A lower F/# results in a blue-shift of the peak wavelength, an increased optical throughput (as characterised by the AUC) and a broadening of the FWHM. These trends are summarised in Table 2.2 and representative examples of the spectral band responses of each sensor are shown in Figs. 2.17, 2.18 and 2.19.

Only the Imec NIR snapshot sensor has a strong F/# dependent spectral baseline (Fig. 2.16d). As for the Imec linescan sensor, the AUC of the baseline decreases (−85%) as the F/# changes from 1.65 to 4.0. It was similarly difficult to comment on the wavelength dependence of the baseline. For quantitative comparison of the in-band spectral response, the data were baseline subtracted.

Strong bimodal responses are observed in the spectral bands of the Imec visible snapshot sensor (Fig. 2.18). The amplitudes of the secondary peaks ranges between 0.10–0.75% of the amplitudes of the primary peaks. 60 ± 14% of the total AUC of

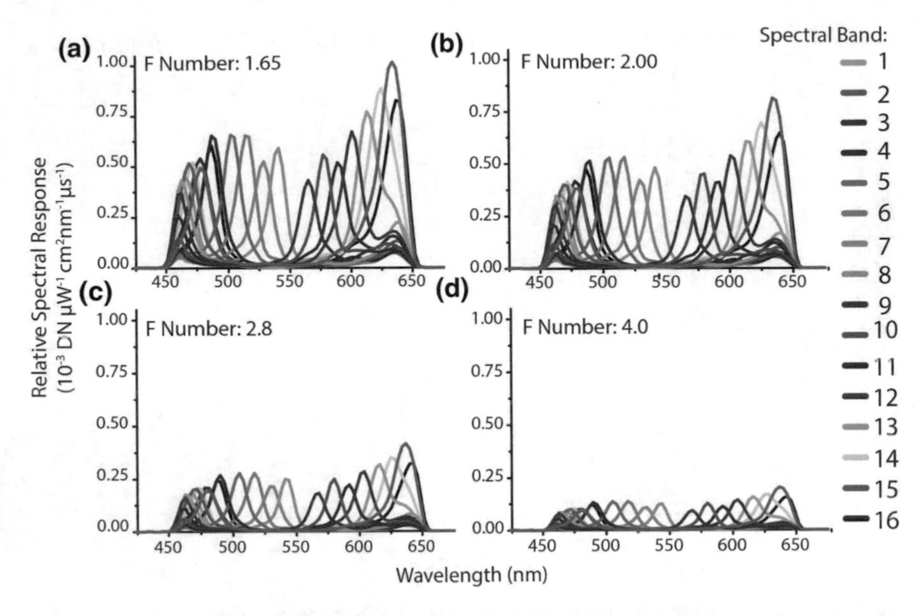

Fig. 2.15 The relative spectral response curves of the Imec visible snapshot sensor for an objective lens F/# of **a** 1.65, **b** 2.0, **c** 2.8 and **d** 4.0

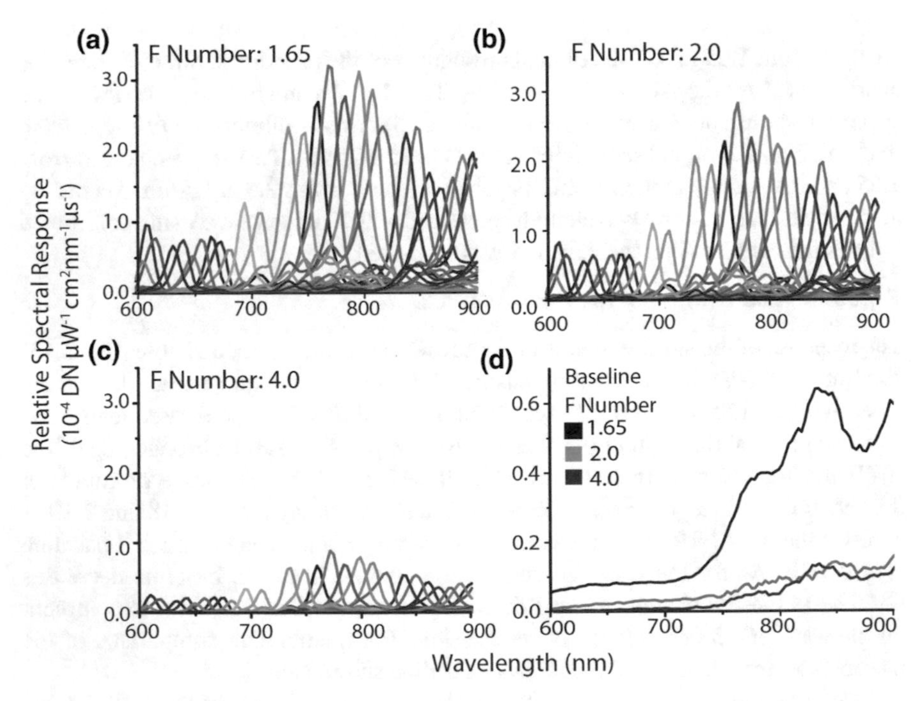

Fig. 2.16 The baseline subtracted relative spectral response curves of the Imec NIR snapshot sensor for an objective lens F/# **a** 1.65, **b** 2.0 and **c** 4.0. **d** The F/# dependent baselines of the relative spectral response curves

Table 2.2 Table summarising the F/# dependence of the spectral transmission characteristics of the primary peak of the spectral response curves of the SILIOS and Imec visible and NIR snapshot sensors

Objective Lens F/# changed from 1.65 to 4.0	SILIOS snapshot sensor	Imec visible snapshot sensor	Imec NIR snapshot sensor
Average shift of peak wavelength	3.4 ± 1.0 nm	3.0 ± 1.1 nm	2.1 ± 1.8 nm
Percentage change of FWHM	$-20 \pm 10\%$	$-13 \pm 6\%$	$-25 \pm 5\%$
Percentage change of AUC within FWHM	$-80 \pm 1\%$	$-82 \pm 1\%$	$-82 \pm 1\%$
Percentage change in ratio of AUC within and outside the FWHM	$-15 \pm 0.8\%$	$-4 \pm 3\%$	$3 \pm 8\%$

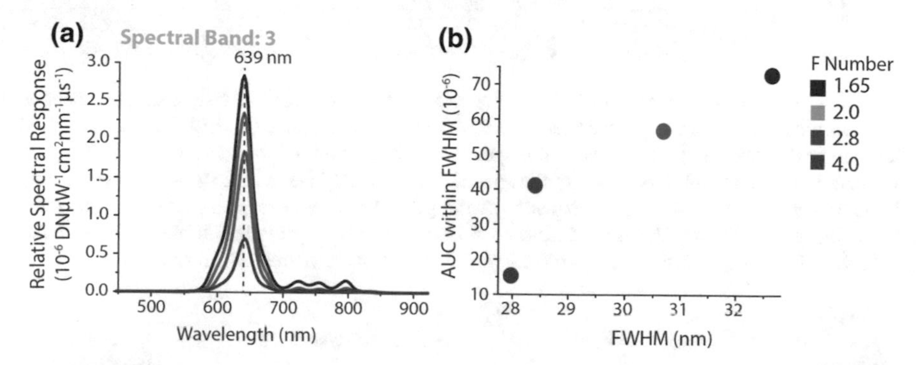

Fig. 2.17 The F/# dependence of a representative spectral band of the SILIOS snapshot sensor. **a** As the F/# of the camera objective lens was decreased, the optical throughput of the filter increased and the peak wavelength of the spectral response was blue shifted (peak wavelength at F/# 1.65 indicated as a reference) **b** At lower F/#s the FWHM of the spectral response became broader as the AUC increased

the spectral bands are contained within the combined FWHM of the two peaks. The spectral response of the bimodal bands was, however, distinctly dominated by the primary peak; $76 \pm 9\%$ of the combined AUCs under the two peaks were contained within the FWHMs of the primary peaks (values extracted at a F/# of 1.65). The amplitude ratio of the two peaks, and the ratio of the AUC within their FWHMs, did not change significantly with F/#. The ratio of the FWHM between the peaks changed by $3 \pm 10\%$ as the F/# was changed from 1.65 to 4.0, the ratio of the AUC contained within the FWHM by $-7 \pm 7\%$, and the total AUC within the two peaks by $6 \pm 7\%$.

Although the supplier provided QE data of the Imec NIR snapshot sensor indicate that certain bands have a bimodal spectral response (Fig. 2.2d), these were not

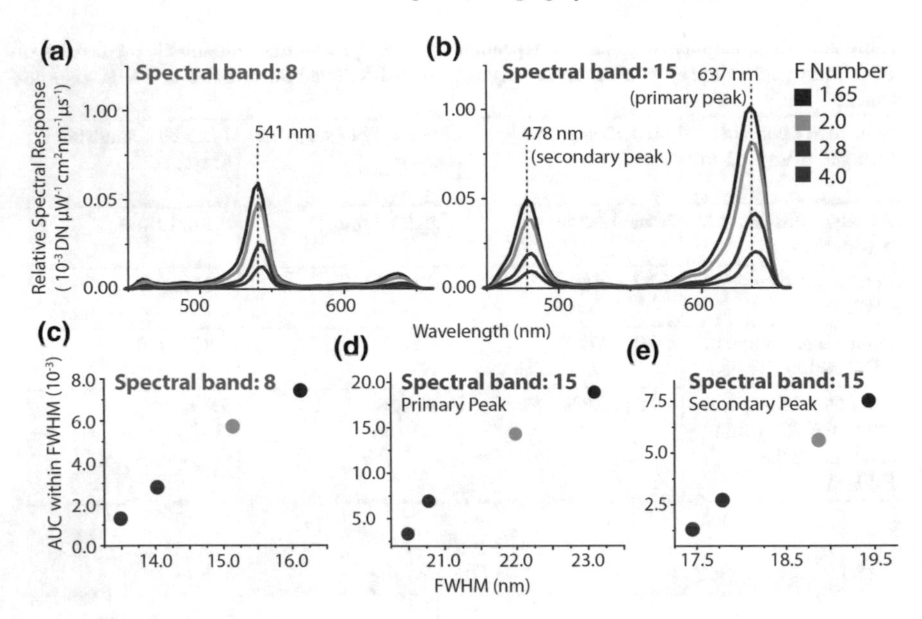

Fig. 2.18 The F/# dependence of representative spectral bands of the Imec visible snapshot sensor. **a** As the F/# of the camera objective lens was decreased, the optical throughput of the filter increased and the peak wavelength of the spectral response was blue shifted (peak wavelength at F/# 1.65 indicated as a reference). **b** A subset of the spectral filters of the Imec visible sensor show a strong bimodal behavior. **c–e** The F/# dependence of the peak wavelength response followed the same trend, independently of whether the spectral band had one or two spectral response peaks; at lower F/#s the FWHM of the spectral response curves became broader as the AUC increased

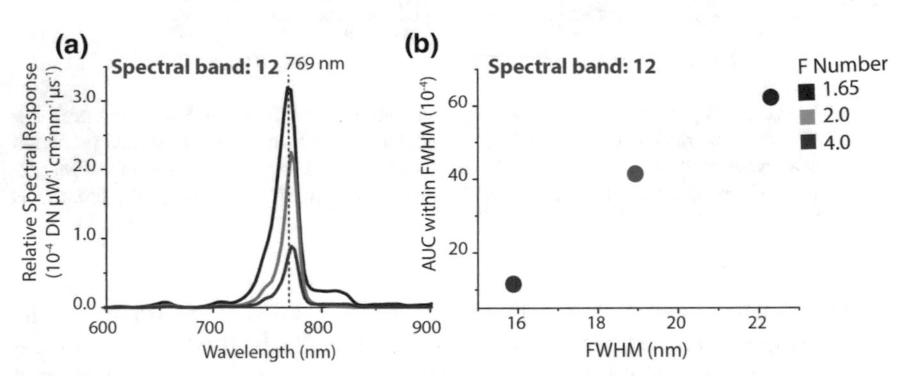

Fig. 2.19 The F/# dependence of a representative baseline subtracted spectral band of the Imec NIR snapshot sensor. **a** As the F/# of the camera objective lens decreased, the optical throughput of the filter increased and the peak wavelength of the spectral responses was blue shifted (peak wavelength at F/# 1.65 indicated as a reference) **b** At lower F/#s the FWHM of the spectral response became broader as the AUC increased

observed within the limited spectral range of the experimental F/# study (380–900 nm vs. 400–1000 nm for the supplier provided data). The analysis was therefore limited to metrics extracted from the primary peaks of the spectral band responses. $29 \pm 10\%$ of the total AUC was contained within the FWHM of the primary peak at F/# 1.65. The ratio of the AUC within the FWHM to the total AUC did not change significantly with F/# ($3 \pm 8\%$).

No bimodal response curves were observed when the full spectral response range of the SILIOS sensor was characterised. At an F/# of 1.65, $53 \pm 7\%$ of the total AUC of the narrowband filters was contained within the FWHM of the peak spectral response. The ratio of the AUC within the FWHM of the primary spectral response curve decreased as the F/# of the objective lens was increased from 1.65 to 4.0 ($-15 \pm 8\%$).

2.4 Discussion and Conclusions

Based on a survey of previously reported spectral imaging systems, the compactness, robustness, and potential for low cost manufacturing make SRDAs particularly well suited for widespread implementation into biomedical and clinical imaging applications. For effective instrumentation integration, the performance characteristics of the sensors do, however, need to be thoroughly understood. Here we performed QE calibrations of SRDAs from two different commercial manufactures—Imec and SILIOS—who use different production processes; the HSI Imec linescan sensor has monolithically integrated spectral filters, whereas the filters of the MSI SILIOS snapshot sensor are deposited on a glass substrate in front of the FPA. We further developed a monochromator-integrating sphere set-up to experimentally study the impact of the camera objective lens' F/# on the spectral response characteristics of the SRDAs. The system was used to quantitatively study the impact of F/# on the spectral response of the Imec linescan, the Imec visible snapshot sensor, the Imec NIR snapshot sensor and the SILIOS snapshot sensor.

Image data from the characterisation showed qualitative differences in the behaviour of the Imec and SILIOS sensors. A periodic noise pattern—not present in spectral band images acquired with the Imec sensors—was observed in spectral band images acquired with the SILIOS snapshot sensor when illuminated with monochromatic light. After further discussions with the sensor manufacturers, the noise pattern was attributed to interference effects between the glass substrate and the CMOS sensors [68].

In this characterisation, the periodic noise pattern did not affect the data since we are only interested in the average spectral response of the sensor. The periodic noise pattern may, however, be problematic when attempting to perform quantitative spectral imaging with the SILIOS sensor. Due to the wavelength dependence of the periodic noise pattern, spectral signatures acquired with the sensor will have a spatial dependence; this could complicate accurate data extraction and subsequent analysis processes. Therefore, the observed interference pattern needs to be quantitatively characterised and corrected to enable effective use of this sensor.

The results from the in-house QE calibrations of the Imec linescan and SILIOS snapshot sensor were well aligned with supplier provided data. We were thus satisfied that we could trust the QE calibration data (for collimated light) provided by both SILIOS and Imec; QE calibrations upon receipt of the Imec snapshot sensors were therefore deemed unnecessary. The in-house QE calibration of the linescan sensor did, however, reveal that pixel rows at the edges of the spectral band had a lower QE(λ) response than the remainder of the pixel rows in the spectral band. Consequently, an 'edge effect' was present in images acquired with the imec linescan sensor; an edge effect that will need to be corrected in software if using the sensor in imaging applications.

An experimental characterisation of the impact of the AOI on the spectral transmission characteristics of the SRDA filters was performed via a monochromator-integrating sphere system and a variable F/# objective lens. The F/# dependence of the Imec linescan and the Imec and SILIOS snapshot sensors were experimentally characterised. The results show that all the SRDAs follow the F/# dependence predicted by theory [60, 62, 64]. When decreasing the F/#, the optical throughput increases, the FWHM broadens, and the peak wavelength of the spectral response curves are blue shifted (see Table 2.2 for quantitative data). In addition, F/# dependent spectral baselines were observed in data acquired with the NIR Imec sensors. As the F/# of the camera objective changed from 4.0 to 1.65, the AUC of the baseline—used as an indicator of the optical throughout—increased by 84 and 85% for the linescan and snapshot sensor respectively. The increasing spectral baselines lead to a lower wavelength selectivity of the spectral bands when imaging at lower F/#. No F/# dependent baselines were observed in data acquired with the imec visible or the SILIOS snapshot sensors. It can therefore be assumed that the baseline behaviour of the visible and NIR Imec sensors may be explained by different material choices in the visible and NIR filter design. The spectral dependence of the baselines and the filter responses show that the choice of accessory optics can impact the imaging performance of SRDAs. The F/# of the camera's objective lens can therefore be used to balance spectral resolution with optical throughput, to best match the requirements of the imaging application. A lower F/# may, for example, be better suited for low-light imaging applications, such as fluorescence imaging.

Note that the presented sensor F/# characterisation does not constitute a full system-level calibration. The current characterisation does not take into account variations in the AOI across the sensor, nor does it produce a per-pixel calibration and correction matrix. We instead characterised the average filter band dependence on F/# to guide the selection of accessory optics used with the SRDAs. Further improvement of the experimental set-up and data analysis protocol would be be required for a full system-level calibration. The illumination uniformity at the output of the integrating sphere would, for example, have to be characterised at each wavelength in the spectral scan. However, for the purpose of studying the general dependence on AOI, characterising the average spectral band response was here considered sufficient. There are also several other applications for which the current dataset may prove beneficial. The F/# dependent spectral response curves could for example be used to improve the initialisation parameters in methods aiming to model full per-pixel correction matrices in software, such as that recently presented by Pichette et al. [33]. Additionally, it would be interesting to compare the results obtained here with CRA modelling of the spectral band response curves for varying objective lens F/#s.

To conclude this chapter, the breadth of technology available for performing spectral imaging has been reviewed, and SRDAs identified as a promising new approach for application in biomedical and clinical imaging. Next, experimental methodology was established to interrogate SRDA performance, in particular assessing the impact of AOI. We have verified supplier provided QE calibration data of SRDAs provided by two commercial sensor manufactures. We have further developed an experimental method to evaluate the impact of the F/# of the camera's objective lens on the imaging performance of SRDAs. The result from this study can guide the selection of accessory optics when imaging with SRDAs, which is the topic of the next two chapters.

References

1. G. Lu, B. Fei, Medical hyperspectral imaging: a review. J. Biomed. Opt. **19**(1), 010901 (2014)
2. D. Waterhouse, A.S. Luthman, S. Bohndiek, Spectral band optimization for multispectral fluorescence imaging. Proc. SPIE Int. Soc. Opt. Eng. **10057**, 1–6 (2017)
3. N. Hagen, M.W. Kudenov, Review of snapshot spectral imaging technologies. Opt. Eng. **52**(9), 090901 (2013)
4. O. Lee et al., Non-invasive assessment of cutaneous wound healing using fluorescent imaging. Skin Res. Technol. **21**(1), 108–113 (2014)
5. S.E. Martinez-Herrera et al., Multispectral endoscopy to identify precancerous lesions in gastric mucosa, *ICISP 2014*. LNCS, vol. 8509 (Springer, Berlin, 2014), pp. 43–51
6. N.T. Clancy et al., Multispectral imaging of organ viability during uterine transplantation surgery in rabbits and sheep. J. Biomed. Opt. **21**(10), 106006 (2016)
7. P.A. Valdes et al., Quantitative, spectrally-resolved intraoperative fluorescence i. Sci. Rep. **2**(798) (2012)
8. R. Leitner, M. De Biasio, T. Arnold, C.V. Dinh, M. Loog, Multi-spectral video endoscopy system for the detection of cancerous tissue. Pattern Recognit. Lett. **34**, 85–93 (2013)
9. P.F. Favreau et al., Excitation-scanning hyperspectral imaging microscope. J. Biomed. Opt. **19**(4), 046010 (2014)
10. A. Rissanen et al., MEMS FPI-based smartphone hyperspectral imager. Proc. SPIE Int. Soc. Opt. Eng. **9855**(985507), 1–16 (2016)
11. J. Herrera-Ramirez, M. Vilaseca, J. Pujol, Portable multispectral imaging system based on light-emitting diodes for spectral recovery from 370 to 1630 nm. J. Appl. Opt. **53**(14), 3131–3141 (2014)
12. M.E. Gosnell et al., Quantitative non-invasive cell characterisation and discrimination based on multispectral autofluorescence features. Sci. Rep. **6**(23453), 1–13 (2016)
13. N. Dimitriadis et al., Spectral and temporal multiplexing for multispectral fluorescence and reflectance imaging using two color sensor. Opt. Express **25**(11), 12812 (2017)
14. C.M. Lee, C.J. Engelbrecht, T.D. Soper, F. Helmchen, E.J. Seibel, Scanning fiber endoscopy with highly flexible, 1 mm catheterscopes for wide-field, full-color imaging. J. Biophotonics **3**(5–6), 385–407 (2010)

15. T.H. Tate et al., Multispectral fluorescence imaging of human ovarian and fallopian tube tissue for early-stage caner detection. J. Biomed. Opt. **21**(5), 014036 (2016)
16. N. Gat, Imaging spectroscopy using tunable filters: a review. Proc. SPIE Int. Soc. Opt. Eng. **4056** (2000)
17. R. Levenson, J. Beechem, G. McNamara, Spectral imaging in preclinical research and clinical pathology. Anal. Cell Pathol. **35**(5–6), 339–361 (2012)
18. Q. Li et al., Review of spectral imaging technology in biomedical engineering: achievements and challenges. J. Biomed. Opt. **18**(10), 100901 (2013)
19. S.A. Mathews, Design and fabrication of a low-cost, multispectral imaging system. J. Appl. Opt. **47**(28), F71–F76 (2008)
20. N. Gupta, P.R. Ashe, S. Tan, Miniature snapshot multispectral imager. J. Opt. Eng. **50**(3), 033203 (2011)
21. B. Geelen, N. Tack, A. Lambrechts, A snapshot multispectral imager with integrated tiled filters and optical duplication. Proc. SPIE Int. Soc. Opt. Eng. **8613**(861314) (2011)
22. M. Denstedt, A. Bjorgan, M. Milanic, L. Lyngsnes-Randeberg, Wavelet based feature extraction and visualization in hyperspectral tissue characterization. Biomed. Opt. Express **5**(12), 4260–4280 (2014)
23. S. Kiyotoki et al., New method for detection of gastric cancer by hyperspectral imaging: a pilot study. J. Biomed. Opt. **18**(2), 026010 (2013)
24. P.J. Cutler et al., Multi-color quantum dot tracking using a high-speed hyperspectral line-scanning microscope. PLoS One **8**(5), e64320 (2013)
25. H.T. Lim, V.M. Murukeshan, A four-dimensional snapshot hyperspectral video-endoscope for bio-imaging applications. Sci. Rep. **6**, 24044 (2016)
26. R.T. Kester, N. Bedard, L. Gao, T.S. Tkaczyk, Real-time snapshot hyperspectral imaging endoscope. J. Biomed. Opt. **16**(5), 056005 (2011)
27. L. Gao, R.T. Kester, N. Hagen, T.S. Tkaczy, Snapshot image mapping spectrometer (IMS) with high sampling density for hyperspectral microscopy. Opt. Express **8**(14), 14340–14344 (2010)
28. L. Gao, R.T. Smith, T.S. Tkaczyk, Snapshot hyperspectral retinal camera with the image mapping spectrometer (IMS). Biomed. Opt. Express **3**(1), 48–54 (2011)
29. L. Gao, L.V. Wang, A review of snapshot multidimensional optical imaging: measuring photon tags in parallel. Phys. Rep. **616**, 1–37 (2016)
30. V.C. Chan, M. Kudenov, C. Liang, P. Zhou, E. Dereniak, Design and application of the snapshot hyperspectral imaging fourier transform (SHIFT) spectropolarimeter for fluorescence imaging. Proc. SPIE Int. Soc. Opt. Eng. **8949**(894903) (2014)
31. Pierre-Jean Lapray, Xingbo Wang, Jean-Baptiste Thomas, Pierre Guton, Multispectral filter arrays: recent advances and practical implementation. Sensors **14**(11), 21626–21659 (2014)
32. A. Lambrechts et al., A CMOS-compatible, integrated approach to hyper- and multispectral imaging. in *2014 IEEE International Electron Devices Meeting, IEDM14* (2014), pp. 261–264
33. J. Pichette, T. Gossen, K. Vunckx, A. Lambrechts, Hyperspectral calibration method for CMOS-based hyperspectral sensors. Proc. SPIE Int. Soc. Opt. Eng. **10110**(101100H), 1–13 (2017)
34. H. Li et al., Snapshot hyperspectral retinal imaging using compact spectral resolving detector array. J. Biophotonics **10**(6–7), 830–839 (2016)
35. J. Pichette et al., Intraoperative video-rate hemodynamic response assessment in human cortex using snapshot hyperspectral optical imaging. Neurophotonics **3**(4), 045003 (2016)
36. T. Sawyer, S.E. Bohndiek, Towards a simulation framework to maximize the resolution of biomedical hyperspectral imaging. Proc. SPIE Int. Soc. Opt. Eng. **10412**(104120C) (2017)
37. C.V. Correra, H. Arguello, G.R. Arce, Snapshot colored compressive spectral imager. J. Opt. Soc. Am. **32**(10), 1754–1763 (2015)
38. C. Yang, V.W. Hou, L.Y. Nelson, R.S. Johnston, D. Melville, E.J. Seibel, Scanning fiber endoscope with multiple fluorescence-reflectance imaging channels for guiding biopsy. Proc. SPIE Int. Soc. Opt. Eng. **8936**, 89360R (2014)
39. S.J. Leavesley et al., Hyperspectral imaging fluorescence excitation scanning for colon cancer detection. J. Biomed. Opt. **21**(10), 104003 (2016)

40. B.P. Joshi, S.J. Miller, C.M. Lee, E.J. Seibel, T.D. Wang, Multispectral endoscopic imaging of colorectal dysplasia in vivo. Gastroenterology **143**(6), 1435–1437 (2012)
41. G. Themelis, J.S. Yoo, K.S. Soh, R. Schulz, V. Ntziachristos, Real-time intraoperative fluorescence imaging system using light-absorption correction. J. Biomed. Opt. **14**(6), 064012 (2009)
42. Y. Lv et al., In vivo simultaneous multispectral fluorescence imaging with spectral multiplexed volume holographic imaging system. J. Biomed. Opt. **2**(6), 060502 (2016)
43. Y. Lv et al., Reduction of blurring in broadband volume holographic imaging using a deconvolution method. Biomed. Opt. Express **7**(8), 3124–3138 (2016)
44. A. Gorman, D.W. Fletcher-Holmes, A.R. Harvey, Generalization of the lyot filter and its application to snapshot spectral imaging. Opt. Express **18**(6), 5602–5608 (2010)
45. L.E. MacKenzie, T.R. Choudhary, A.I. McNaught, A.R. Harvey, In vivo oxiometry of human bulbar conjunctival and episcleral microvasculature using snapshot multispectral imaging. Exp. Eye Res. **149**, 48–58 (2016)
46. M.W. Kudenov, M.E.L. Jungwirth, E.L. Dereniak, G.R. Gerhart, White-light sagnac interferometer for snapshot multispectral imaging. Appl. Opt. **49**(1), 4067 (2010)
47. W.R. Johnson, D.W. Wilson, W. Fink, M. Humayun, G. Bearman, Snapshot hyperspectral imaging in opthalmology. J. Biomed. Opt. **12**(1), 014036 (2014)
48. S.P. Burgos, S. Yokogawa, H.A. Atwater, Color imaging via nearest neighbor hole coupling in plasmonic color filters integrated onto a complementary metal-oxide semiconductor image sensor. ACS Nano **7**(11), 10038–10047 (2013)
49. M. Najiminaini, F. Vasefi, B. Kaminska, J.J.L. Carson, Nanohole-array-based device for 2d snapshot multispectral imaging. Sci. Rep. **3**(2589), 1–7 (2013)
50. H. Park, K.B. Crozier, Multispectral imaging with vertical silicon nanowires. Sci. Rep. **2460**(3), 1–6 (2013)
51. L. Duempelmann, B. Gallinet, L. Novotny, Multispectral imaging with tunable plasmonic filters. ACS Photonics **4**(2), 236–241 (2017)
52. S. Pellicori, Coating processes evolve: satisfying special optical requirements. Materion: Technical paper (2017), https://materion.com/-/media/files/pdfs/advanced-materials-group/ac/ac-newsletter-article-pdfs/coating-processes-evolve_technical-paper.pdf. Accessed 25 Apr 2017
53. I.G.E. Renhorn, D. Bergström, J. Hedborg, D. Letalick, S. Möller, High spatial resolution hyperspectral camera based on a linear variable filter. Opt. Eng. **55**(11), 114105 (2016)
54. Pixelteq, Micro-patterned optical filters. January, 2016
55. J.B. Thomas, P.J. Lapray, P. Gouton, C. Clerc, Spectral characterization of a prototype SFA camera for joint visible and NIR acquisition. Sensors **16**(7), 1–19 (2016)
56. P. Gonzalez et al., A novel CMOS-compatible, monolithically integrated line-scan hyperspectral imager covering the VIS-NIR range. Proc. SPIE Int. Soc. Opt. Eng. **9855**(98550N), 236–241 (2016)
57. O. Pust, Innovative filter solutions for hyperspectral imaging (2016), www.optik-photonik.de
58. N. Tack, A. Lambrechts, P. Soussan, L. Haspeslagh, A compact, high-speed, and low cost hyperspectral imager. Proc. SPIE Int. Soc. Opt. Eng. **8266**, 82660Q (2012)
59. A. Kristensen et al., Plasmonic colour generation. Nat. Rev. Mat. **2**(16088), 1–14 (2016)
60. L. Frey, L. Masarotto, M. Armand, M.L. Charles, O. Lartigue, Multispectral interference filter arrays with compensation of angular dependence or extended spectral range. Opt. Express **23**(9), 11799–11812 (2015)
61. Q. Chen, X. Hu, L. Wen, Y. Yu, D.R.S. Cumming, Nanophotonic image sensors. Small **12**(36), 4922–4935 (2016)
62. H.A. Macleod, Chapter 6: Edge filters and Chapter 7: Band-pass filters, *Thin Film Optical Filters*, 3rd edn. (IoP, Bristol, 2000)
63. S.V. Kartalopoulus, Chapter 3: Introduction to optical components, *Introduction to DWDM Technology* (Wiley-IEEE Press, New York, 2000)
64. P. Agrawal et al., Characterization of VNIR hyperspectral sensors with monolithically integrated optical filters, in *Proceedings of IS&T International Symposium on Electronic Imaging, 2016* (2016)

65. M. Jayapala et al., Monolithic integration of flexible spectral filters with CMOS image sensors at wafer level for low cost hyperspectral imaging in international image sensor workshop, in *Snowbird* (2013)
66. S. Pellicori, The evolution of coating requirements: meeting special optical requirements. Materion: Technical paper (2017), https://materion.com/-/media/files/pdfs/precision-optics/technical-papers/evolutionofcoatingprocesses_mpo.pdf. Accessed 25 Apr 2017
67. SILIOS Technologies, CS filters for multispectral imaging (2017), http://www.silios.com/cs-filters-for-multispectral-imaging. Accessed 26 Apr 2017
68. S. Tisserand, Email Communication to S.E. Bohndiek, A.S. Luthman and D. Waterhouse. Private Communication, 21 July 2016
69. J. James, *Photon Transfer: DN->λ* (SPIE, Bellingham, 2007)
70. EMVA standard 1288: standard for characterization of image sensors and cameras, release 3.0. Technical report, European Machine Vision Association, November 2010
71. A.J.P. Theuwissen, CMOS image sensors: state-of-the-art. in *2007 33rd European Solid State Circuits Conference, ESSCIRC 2007*, vol. 52 (2008), pp. 1401–1406
72. S.E. Bohndiek, Comparison of methods for estimating the conversion gain of CMOS active pixel sensors. IEEE Sens. J. **8**(10), 1734–1744 (2008)
73. A. Theuwissen, How to measure "photon transfer curve" (1)?. Harvest Imaging Blog (2012), http://harvestimaging.com/blog/?p=1034. Accessed Aug 2012
74. Fabry-Perot: wider FWHM means reduced attenuation/optical density. Technical report, Imec, 2015
75. SILIOS, CMS: multi-spectral camera. Specification Sheet
76. MATLAB. pdist (2017), https://uk.mathworks.com/help/stats/pdist.html. Accessed 17 Aug 2017
77. E. Hecht, *Chapter 9: Interference, Optics*, 4th edn. (Pearson Education Limited, Essex, 2014)

Chapter 3
Wide-Field fHSI with a Linescan SRDA

Clinical spectral imaging has the potential to improve diagnosis and guide treatment [1]. Tissue contrast may be extracted from the reflectance, absorption and fluorescence spectra. Spectral imaging may, for example, be used to extract a multiplexed signal from endogenous tissue chromophores or from exogenously administered fluorescent contrast agents, via separation of their absorption and/or emission spectra.

The combined development of targeted fluorescent contrast agents and spectral imaging instrumentation advances has enabled multiplexed fluorescence imaging; the colouring of tissue with a cocktail of contrast agents, targeting several molecular pathways [2]. Molecular fluorescence imaging is now emerging as a novel intraoperative and endoscopic imaging method, for which the ability to resolve several different fluorescent dyes in real-time present several advantages [3, 4]. The use of MSI/HSI systems may enable multiplexed fluorescence imaging. In Chap. 2, SRDAs were identified as a promising new approach to acquire spectral image data in biomedical and clinical applications. This chapter introduces a first proof-of-concept system for multiplexed fHSI using a SRDA. Here we integrated the Imec linescan sensor (CMV2K-LS600-975-2.4.8.4; Imec) into a reflectance-based wide-field imaging system to demonstrate HSI of fluorescent dyes in well plates, tissue mimicking phantoms and in vivo. Up to seven fluorescent dyes—excited by light from a LED ring—were simultaneously imaged, and spectrally unmixed, using HSI data acquired with the Imec linescan sensor suspended over a sample stage. Whereas the system has intrinsic value as a fHSI imaging platform, it also allowed us to explore solutions to software and hardware challenges associated with integrating a SRDA into a fHSI system.

This chapter details the current uses of biomedical and clinical multiplexed fluorescence imaging and discusses the technical limitations preventing widespread clinical use of this technique. The fHSI system design and initial multiplexed in vitro and in vivo fluorescence imaging results are presented. The developed fHSI system was initially characterised, followed by an investigation of the efficacy of background removal methods, and finally the imaging performance of the system

A. S. Luthman, *Spectrally Resolved Detector Arrays for Multiplexed Biomedical Fluorescence Imaging*, Springer Theses,
https://doi.org/10.1007/978-3-319-98255-7_3

was evaluated by imaging fluorescent dyes on a range of backgrounds. Here we demonstrate unmixing of seven fluorescent dyes in solution and at least four dyes in tissue mimicking phantoms and also following subcutaneous injection in nude mice.

Parts of the work presented in this chapter have been previously published in the *Journal of Biophotonics* [5].

3.1 Literature Review

3.1.1 Multiplexed Fluorescence Imaging for Biomedical Applications

With spectral imaging instrumentation advances the spectral contrast of tissue has been increasingly explored for biomedical and clinical imaging. Up to date, a majority of the literature has focused on intrinsic spectral contrast, extracted from tissue via reflectance imaging in the visible and NIR spectral region [6]. In the visible spectral region, intrinsic tissue contrast is high due to AF and differential absorption from endogenous chromophores, such as oxy/deoxygenated haemoglobin [2, 7]. The diagnostic capability of intrinsic tissue contrast in the visible spectral region has been shown to allow the assessment of diabetic foot ulcers [6]; vascular conditions [6]; tissue perfusion [8, 9]; melanoma screening [10]; wound healing [11, 12]; ophthalmology [13–15]; cancer diagnostics [16–18]; tumour resection margin determination [19] and surgical guidance [20, 21]. In comparison to the visible spectral region, tissue absorption, scattering and AF is relatively low within the NIR [2, 22, 23]. This increases the penetration depth of light in tissue and allows imaging of extrinsic fluorescent contrast agents with high SBRs [22].

Preclinical optical imaging mainly rely on exogenously administered fluorescent contrast; either via transfection of fluorescent proteins, or the injection of fluorescent contrast agents [24]. Exogenous fluorescence is preferable to intrinsic fluorescence as it provides a more specific signal and enables interrogation of a much wider range of biological processes than obtainable from intrinsic fluorophores [7, 24, 25]. In vivo molecular imaging can, for example, be achieved using fluorescent contrast agents that binds to cell surface receptors for which disease specific changes can be detected via over, or under, expression of the receptor in response to disease [1]. The increased availability of extrinsic fluorophores operating within the NIR spectral window also enables in vivo imaging of extrinsic fluorescent contrast agents with high contrast and SBR [2, 22].

Spectral imaging has the potential to allow multiplexed fluorescence imaging of several fluorescent contrast agents via spectral separation of their emission spectra. Multiplexed fluorescence imaging enables the simultaneous imaging of a “cocktail of targeting fluorophores” for concurrent imaging of different molecular pathways [1, 3, 4]. Simultaneous readout from several biomarkers may help overcome one of the main challenges of biomedical fluorescence imaging; extracting valuable biomedical information from tissues with heterogeneous target expression [3, 4]. On a larger scale, multiplexed fluorescence imaging can also aid the distinction between different tissue structures [26]. For example, Hyun et al. [26] used a 700 nm and a 800 nm fluorophore, with organ specific retention, to distinguish between the parathyroid and thyroid glands during head and neck surgery in pigs.

In addition to revealing different tissue structures and molecular processes, multiplexed fluorescence imaging may also improve imaging SBR. The SBR can, for example, be improved via the paired probe method [1]. The paired probe method promotes the detection of a targeted contrast agent by the inclusion of a non-targeted fluorescent contrast agent, used as a control for non-specific binding [1]. Multiplexed fluorescence imaging of a dual wavelength fluorescent probe, with distinct visible and NIR excitation and emission spectra, has also been shown to enable depth estimation of the fluorescence signal origin in vivo [27]. Such depth resolved imaging has the potential to reveal the underlying biological structure of tissues and may allow for improved diagnosis and possibly improved resection of tumours [27]. Indeed, the ability to detect multiple fluorescent labels targeted to different biological pathways has been shown highly advantageous across a range of biomedical imaging applications; from studying complex molecular interactions in cells at super-resolution level [28], through monitoring contrast agent biodistribution in small animals [22, 24, 29], to diagnosing, characterising and resecting cancers in humans [1, 22].

The utility of multiplexed fluorescence imaging has already been well established in microscopy, and consequently all major microscope manufacturers now carry multispectral systems [30]. Multispectral small animal imaging chambers for pre-clinical imaging are also commercially available [31–34]. Clinical applications of multiplexed fluorescence imaging have, however, been limited by the lack of Federal Drug Administration (FDA) or European Medicine Agency (EMA) approved NIR fluorescent contrast agents [22]. Currently no targeted NIR fluorescent dyes have gained full regulatory approval for clinical use [22]. A majority of clinical fluorescence imaging studies have therefore been conducted with the regulatory approved non-targeted fluorescent dyes indocyanine green (ICG), methylene blue, or with 5-aminolevulinic acid (5-ALA) induced protoporphyrin IX (PpIX) fluorescence [22]. Indeed, the application of non-targeted fluorescent contrast agents has improved the diagnostic performance of imaging [35]. Clinical multiplexed fluorescence imaging using targeted fluorescence contrast agents promises to push the diagnostic imaging performance further.

Within the last 10 years, several targeted NIR fluorescent contrast agents have been developed. The clinical applications of multiplexed fluorescence imaging have, however, been held back by the lengthy and costly process of obtaining regulatory approval for new contrast agents [1]. The lack of suitable imaging platforms has also partly impeded the progress of regulatory approved novel contrast agents, since naturally the clinical performance of a contrast agent depends on the combined performance of the contrast agent and the imaging system used [1, 35]. Consequently the clinical approach for novel fluorescent contrast agents has therefore often been to pair the agent to a specific model of imaging device [1, 35]. This approach may ultimately facilitate the acceptance of flexible imaging devices that are sensitive to a wide range of different fluorescent emissions with the ability to separate signals from multiple contrast agents; such as MSI and HSI systems.

Further development of MSI and HSI imaging systems optimised for multiplexed clinical fluorescence imaging applications is therefore required. The currently commercially available microscopy and preclinical MSI and HSI imaging systems relay on filter cubes and tunable spectral filters [30–34]; thus suffering from the limitations discussed in Chap. 2. Optical filters make the systems bulky and require sequential acquisition of spectral image data limiting their use in clinical imaging applications, such as intraoperative and endoscopic imaging where data typically need to be acquired at video rate. In contrast, snapshot SRDAs enables the acquisition of image data in several spectral bands at video rate without the need for additional spectral filters. Clinical instrumentation often need to be transportable and robust. The compact and robust nature of SRDAs therefore makes them well suited to the clinical environment. Although SRDAs have not been previously used for multiplexed fluorescence imaging, early applications of snapshot SRDAs for spectral white-light reflectance imaging have indeed shown promise both for neurosurgery [20] and ophthalmology [14].

The Imec linescan sensor was here used to create a reflectance-based wide-field imaging system in order to provide a first proof-of-concept demonstration of fHSI with a SRDA. While the SRDA filters can be deposited pixel-wise for real-time multispectral snapshot imaging, we used a sensor with row-wise filters to perform linescan HSI in order to maintain high spatial and spectral resolution in this initial technology evaluation. Following the sensor characterisation detailed in Chap. 2, the wide-field reflectance based fHSI system allowed us to explore the challenges associated with system integration of SRDAs, and evaluate the imaging performance of the Imec linescan sensor on a wide range of fluorescent samples.

To provide an initial proof-of-concept for fHSI relevant to clinical applications, we created a simple reflectance-based demonstrator system. We here used LEDs for fluorescence excitation to mimic the typical sample and illumination geometry encountered in clinical applications. This allowed us to specifically capture and explore the hardware and software challenges of performing fHSI without adding all of the complexities that would be associated with a fully functional endoscopic or intraoperative system. Here we explored and evaluated; the impact of F/# for fHSI, HSI sensor fluorescence response linearity, spectral unmixing precision, and delineation of reflectance and fluorescence light. We showed that the wide-field fHSI system, based on the linescan SRDA, can resolve at least 7 fluorescent dyes in solution and at least 4 dyes in tissue mimicking phantoms as well as in vivo in small animals. These results show promise for future clinical fluorescence imaging applications of SRDAs.

3.1.2 *Spectral Unmixing and Reflectance Background Removal*

A 3D data cube arising from spectral imaging contains spatial and spectral information, such that each image pixel has an associated spectrum. To derive useful information, it is necessary to relate the spectral data to the chemical composition of the sample. Such relationships can be established via spectral unmixing. The end goal of spectral unmixing is to extract the abundances of the individual components of a sample by separating the reflectance, absorption or fluorescence spectra of its constituents. The spectra of the individual components are referred to as endmembers, and spectral unmixing can involve estimating the number of endmembers, their spectral signatures and their abundances at each spectral pixel [36].

In general, spectral unmixing techniques are divided into supervised and unsupervised methods. Whereas supervised unmixing techniques assumes knowledge of the number of endmembers and their spectral profile, unsupervised spectral unmixing additionally involves the determination of the number of endmembers and their spectral signatures from the acquired data [37]. Unsupervised spectral unmixing methods may be broadly categorised as statistical or geometrical methods [38]. Whereas geometrical methods rely on the identification of pure pixel spectra, statistical algorithms identify endmembers from natural partitions in the data, based on the distinct statistical behaviour of the spectral data [37]. Many methods which determine the number of endmembers and their spectral profile, also simultaneously approximate their relative abundances [37]. For many spectral unmixing problems, the endmembers will however be known a priori; the unmixing problem is then reduced to an inversion problem aiming to determine the relative endmember abundances from the acquired data. Although several different inversion algorithms exist, a majority of the available and most commonly applied methods are based on least squares spectral unmixing [38, 39].

Spectral unmixing methods are applied in a wide range of fields, ranging from remote sensing [36] to biomedical imaging [39]. Consequently, a large number of spectral unmixing methods and techniques have been developed within the last decade across a range of disciplines, often independently of each other [40]. Reviewing these unmixing methods is however beyond the scope of this thesis, since the classical supervised method of linear least squares spectral unmixing was showed to perform well for the acquired data.

Unsupervised spectral unmixing is well suited for in vivo reflectance imaging [41], where the identity of the endogenous chromophores present in tissue is typically not known a priori [41]. For multiplexed fluorescence imaging, where the spectral signature of the fluorescence is typically known a priori, the complexity of the spectral unmixing method may however be reduced by the use of a library of reference endmembers.

In this thesis, a linear mixing model of the endmembers was assumed. A linear mixing model assumes that each endmember contributes to the optical signal according to its fractional abundance in the pixel, without taking into account scattering of the photons, or interactions between the different endmembers [36]. The pixel spectra at a position n (y_n), may therefore be represented as a linear combination of the endmembers (a_i) weighted according to their fractional abundances (s_i) at the pixel position, according to;

$$y_n = \sum_{i=1}^{N} a_i s_i[n] + w_n, \tag{3.1}$$

where w_n denotes an additive perturbation from noise or modelling errors, and N denotes the number of endmembers [37]. Despite the naivety of the linear mixing assumption, this model is commonly considered an acceptable approximation for spectral unmixing [36, 37] and is frequently applied within biomedical imaging [39]. It is typically solved by constrained linear least squares;

$$\hat{\mathbf{s}}[n] = arg \min_{s[n]\in S} ||\mathbf{y}_n - \mathbf{A}\mathbf{s}[n]||^2, \tag{3.2}$$

where S denotes the feasible set of endmembers [37]. Equation 3.2 is solved for the abundances of the endmembers at each pixel ($\mathbf{s}[n]$) by iteratively minimising the squared error between the data ($\mathbf{y}_n$) and the estimation of the endmembers ($\mathbf{A}$) [37]. For realistic extraction of the endmembers, a non-negativity least squares (NNLS) constraint is commonly included in the iterative process. Equation 3.2 is therefore optimised based on the condition that $s_i \geq 0$ [36, 37].

For the reflectance based fHSI system developed here, the strongest optical signal measured in the spectral signatures is the reflectance from the illuminating LEDs rather than the fluorescent dye endmembers. We therefore tested two software background removal methods (which exploit the spectral dimension of the information collected in the HSI cube) and compared the result to the inclusion of crossed linear polarisers in the hardware (which remove only surface specular reflections). The advantage of performing background removal in the software, is the potential to extract information from the diffuse reflectance light in a clinical scenario, assuming we are able to achieve a comparable performance for reflectance removal as in hardware. The performance of a 'brute force' software approach—mimicking the commonly used hardware action of applying optical bandpass filters to block the reflectance light—was compared to a statistical background removal method.

The spectral dimension of the acquired data was exploited by implementing statistical background removal via orthogonal subspace projection (OSP). OSP is a widely applied statistical background removal method [42], which we applied to remove the impact of the reflectance light. The basic concept of OSP is to project the endmember spectra to a basis subset which is orthogonal to the background reflectance signal [43]. A matrix of the basis spectra of the background ($\mathbf{V}$) can be straightforwardly extracted, by performing principal component analysis (PCA) on a separately acquired spectral data cube which does not contain any of the fluorescent dye endmembers [43]. The projections of $\mathbf{V}$ on the fluorescent dye endmembers ($\mathbf{A}$)

were then subtracted to produce background removed endmembers (**r**), according to;

$$\mathbf{r} = \mathbf{A} - \mathbf{A}\mathbf{V}^T\mathbf{V}. \tag{3.3}$$

This method assumes that at least some of the endmembers are orthogonal to the background subspace [42]. OSP was thus considered well-suited to the fHSI system since the LED illumination was specifically selected to minimise spectral overlap with the fluorescent dye emissions.

Note that whereas the literature contains a multitude of spectral unmixing and background removal algorithms, we have focused our imaging performance comparison on three representative background removal methods whilst consistently implementing a common spectral unmixing algorithm. This allows us to explore which category of background removal methods may be best suited for reflectance based fHSI.

3.2 Experimental Methods

3.2.1 System Design, Calibration and Characterisation

The Imec linescan HSI camera was mounted onto a strong aluminum frame (XT95-500, XT95P3, XT95P12/M and XT95P11/M; Thorlabs) above an automated micrometer stage (LTS150; Thorlabs) (Fig. 3.1a), using a custom adapter designed to minimise mechanical vibrations. A wide-field objective lens with variable aperture (35 mm VIS-NIR Compact Fixed Focal Length Lens, Edmund Optics) and the IR long-pass filter, previously included in the F/# characterisation of the Imec linescan sensor (Sect. 2.2.3), was used to control the light reaching the sensor. The system was enclosed in a light-tight box.

Spatial Scanning of Sample

The sample was scanned under the HSI camera, with a single image acquired at each scan position. Each image from the HSI camera has dimensions of x–λ, while the linear scanning adds the y dimension. The HSI data cube was reconstructed off-line in MATLAB® 2013/2015 (Fig. 3.1b) after the spectral scan by stitching together the spectral data from each y step to produce an x–y–λ matrix ($2048 \times 576 \times 72 =$ 84.9 MegaPixels; x–y pixel dimension 34.2 μm in the field of view (FOV)). Linear scanning required an acquisition time of up to 4 min per HSI cube, but allowed us to maximise the spectral and spatial information available from the SRDA.

The spatial scanning was performed with an automated micrometer stage (LTS150; Thorlabs) controlled by a LabVIEW® script. One image was acquired at each step of the micrometer stage during the spatial scan; in total 143 consecutive image frames were acquired to produce an HSI data cube of 72 spectral band images. To allow

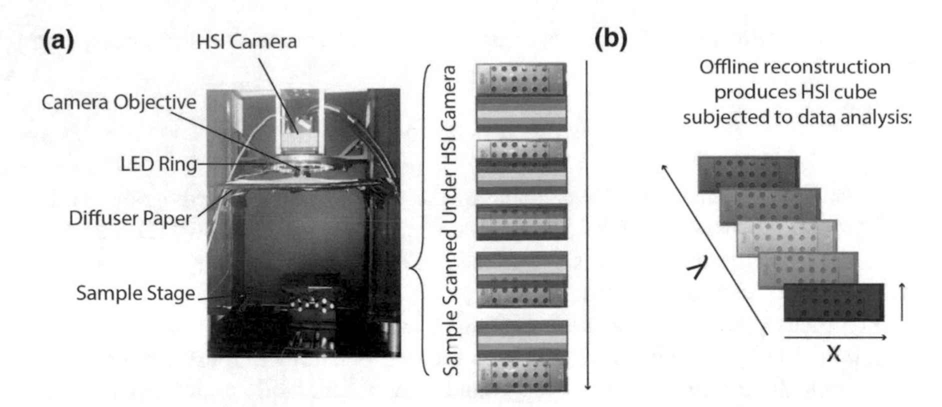

Fig. 3.1 **a** A wide-field reflectance fHSI imaging system was realised by suspending the Imec linescan sensor above a sample platform mounted on a micrometer translation stage. **b** The sample is scanned under the HSI camera, and the HSI data cube reconstructed off-line by stitching together the spectral band data from each scan position

for adjustment of the vertical distance between the sample platform and the camera, the sample platform was placed on a lab-jack (L200/M; Thorlabs). The height of the sample platform and the focus of the objective lens were adjusted such that the spectral FOV was 70 mm × 20mm. The lab jack was placed on a manually controllable micrometer stage (PT1/M; Thorlabs) to allow for fine alignment of the sample position in the direction orthogonal to the spatial scanning.

As the size of the filter bands in the FOV depends on the focal length and the objective lens, the step size of the micrometer stage had to be calibrated for the specific imaging set-up. A grid system was used to ensure accurate alignment of the sample stage and to set the appropriate micrometer stage step-size. A grid pattern, showing the position of the spectral filter bands of the SRDA, was overlaid on the video rate feed from the camera via the LabVIEW® software. The same grid was also rescaled to the size of the camera's FOV and printed on overhead paper. The printed grid was placed on a uniformly spectrally reflective target (Sphere Optics Lambertian White Screen; SG3151-0) on top of the sample stage; the printed grid pattern was illuminated with a broadband halogen lamp, to ensure that the grid was fully visible in the video rate feed from the camera. The printed grid was then manually aligned with the grid overlaid on the video rate feed from the camera in order to calibrate the initial position, final position and step size of the micrometer stage scan range.

Illumination for Wide-Field f-HSI System

HSI traditionally uses either broadband white light or narrowband filters for illumination [21]. Here, an intermediate approach was taken using a LED ring composed of 3 LED colours for fluorescent excitation; one outside of the spectral response range of the HSI camera at 590 nm and two within the range at 660 nm and 732 nm (LZ1-10A100, LZ1-10R200, LZ1-10R30; MCPCB mounted from LED Engine).

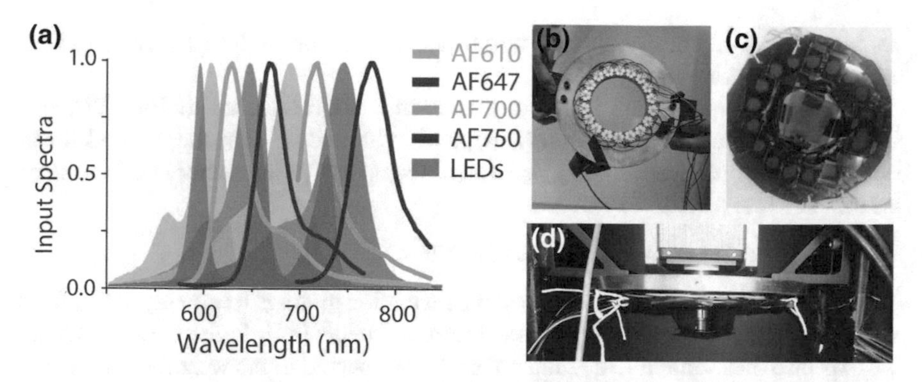

Fig. 3.2 **a** LEDs, matched to the absorption spectra of the four main fluorescent dyes used in this study (AlexaFluor 610, 647, 700, 750), were used for sample illumination. The normalised total LED emission spectrum and the absorption spectra of the dyes are here shown by shaded colour plots, whilst the fluorescence emission spectra of the dyes are indicated by corresponding line plots. **b** LEDs were placed equidistantly around the internal rim of a custom copper ring, later mounted below the HSI camera, to provide sample illumination. To compare hardware specular reflection removal with reflectance removal methods implemented in software, data were also acquired with crossed linear polarisers (**c**) orthogonally placed in front of the camera objective lens and LED ring (**d**)

The LED outputs were matched to the absorption of the main fluorescent contrast agents (AlexaFluor (AF) 610, 647, 700 and 750) used in this study (Fig. 3.2a).

The LED ring consists of a custom copper ring of 50 mm internal radius with six LEDs of each colour connected in series placed equidistant around the internal rim of the ring (Fig. 3.2b). LEDs were chosen to allow flexible 'plug and play' exchange of excitation wavelengths, with easy removal and replacement of LEDs allowed by attaching the LEDs to the copper ring with double sided heat-conducting tape (BP100-0.005-00-00-12; RS). A photographic diffuser paper (Lee White Diffusion; Calumet Photographic) was placed underneath the LED ring to further improve the spatial uniformity of the illumination.

For imaging, the power density of each LED colour was set to $100\,\mu\text{Wcm}^{-2}$ via separate current controllers (LEDDD1B; Thorlabs) and measured with a calibrated optical power meter (1916-R; Newport) centrally placed in the sample area. The LED intensity drift over a time period of 15 min was assessed by recording the light intensity in 30 sec intervals with the optical power meter. Emission spectra of the LEDs were also independently verified using a commercial spectrometer (AvaSpec-ULS2048-USB2-FCDC; Avantes). The normalised spectrum from the LEDs is shown in Fig. 3.2a. Throughout the data acquisition, light from all LEDs simultaneously illuminated the full FOV.

To compare hardware specular reflection removal with reflectance removal methods implemented in software, data acquired with and without crossed linear polarisers (Fig. 3.2c, ALPVISE2X2; Thorlabs), orthogonally placed in front of the camera objective lens and LED ring (Fig. 3.2d), were compared. The crossed linear polariser set-up was designed and assembled by Mr. Sebastian Dumitru (Master Student, University of Cambridge, UK).

The spatial illumination uniformity of each LED colour, set to $100\,\mu W/cm^{-2}$ in the sample plane, was separately mapped, with and without the crossed linear polarisers, by shifting the x and y position of the optical power meter in 0.5 cm steps across the sample area.

Evaluation of Keystone and Spectral Smile

Traditional linescan HSI sensors based on diffraction gratings frequently suffer from aberrations such as spectral smile and keystone effects [44]. In the current system, we expected these effects to be minimal since the spectral filters were monolithically integrated on the sensor. We did however characterise both keystone and smile for experimental completeness.

Spectral smile is the shift in the peak wavelength for pixels along a row orthogonal to the scanning direction [44]. This typically produces a characteristic smile like curve when plotting the peak wavelength of the spectral response curves of pixels in the same row [44]. We did expect to see some variations in the peak wavelength across the pixel rows of the SRDA, since the AOI of the light is expected to vary slightly across the sensor [45]. To qualitatively investigate the smile effect, we therefore illuminated and imaged a uniform reflectance target (Sphere Optics Lambertian White Screen; SG3151-0) with the 660 nm LEDs of the LED ring, following the standard image acquisition and processing protocol later detailed in Sect. 3.2.2. The camera objective lens F/# was set to 1.65, as this was experimentally determined as the most suitable F/# for the low light application of fluorescence imaging, as described in Sect. 3.3.2.

Keystone aberrations typically arise from a wavelength dependent shift of the optical magnification of the HSI system [44]. This is manifested as a spatial shift of pixels with the same peak wavelength response [44]. To investigate keystone aberrations, a grid pattern was printed on overhead paper and placed on top of the reflectance target. The sample was illuminated with a broadband halogen light source (OSL2 with OSL2B2 bulb; Thorlabs) to obtain reflectance images in all spectral bands. Data were acquired with a camera objective lens F/# of 1.65, following the standard image acquisition and pre-processing protocols, detailed in Sect. 3.2.2.

3.2.2 Imaging Protocol and Hyperspectral Data Pre-processing

For each experiment, a dark, a 'flat' and a 'reference' HSI data cube were acquired additionally to the fHSI data; here the 'flat' and 'reference' data cube refers respectively to a HSI data cube of the sample prior to the addition of fluorescent dyes and a HSI data cube with the fluorescent dyes placed in known spatial positions for extraction of endmembers for spectral unmixing. Whereas the fluorescent dye endmembers may be calculated from supplier provided spectra, we here chose to measure their

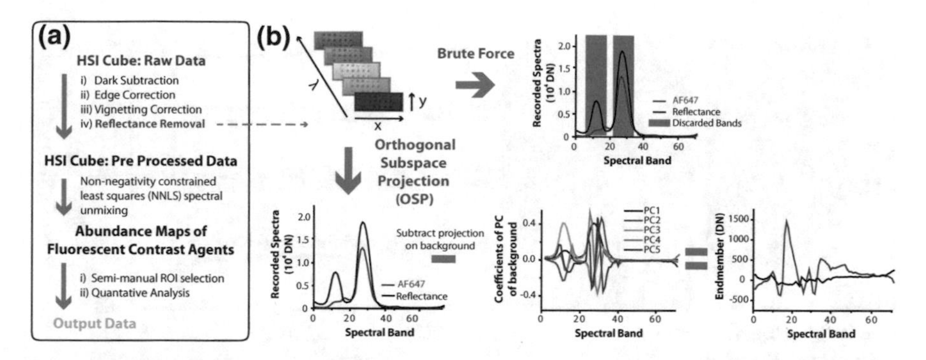

Fig. 3.3 Data processing steps required to extract fluorescent dye abundance maps from the acquired fHSI data cubes. **a** The raw data were initially pre-processed by dark subtraction, edge correction, (to correct for position mismatches between the filter bands and pixels edges) and vignetting correction. **b** Reflectance removal methods were applied in software using 'brute force' background removal and OSP. For 'brute force' background removal, spectral bands containing reflectance peaks were simply discarded from further analysis. OSP exploits the spectral component of the fHSI data cube. The projection of the fluorescence data on the principal components of a 'flat' HSI data cube (representing the background) are subtracted to separate the fluorescence signal from the reflectance light prior to spectral unmixing

fluorescence spectra in situ as it is known that the excitation and emission properties of fluorescent molecules can be significantly affected by their environment [46]. Direct measurement of the fluorescent dye spectra has the added benefit of directly taking into account the modulation of the spectral signatures by the system response.

The standard image pre-processing steps included: dark subtraction; edge correction and a spatial vignetting correction (Fig. 3.3a). Two software reflectance removal methods (Fig. 3.3b) were also separately implemented in software, statistical background removal via OSP and a 'brute force' method by straightforward omission of the spectral bands dominated by the reflectance signal from the LEDs. Spectral unmixing of reflectance and non-reflectance removed data were subsequently performed by NNLS unmixing. Further details on each pre-processing step are given in the following sections.

Edge Correction

Edge corrections were performed to correct for the lower QE of pixels rows positioned at the edges of the spectral bands. The lower QE was first identified in the sensor calibration of the Imec linescan sensor (Sect. 2.3.1), but can also be observed as a line artefact in the acquired spectral band images (Fig. 3.4a). To remove the line artefact, pixels at the edge of the spectral bands were replaced with the average intensity response of their adjacent pixels, representing a simple form of spatial denoising (Fig. 3.4b, c).

Vignetting Correction

Spatial vignetting corrections are typically performed via flat fielding. Flat fielding corrects uneven illumination and spectral variability of the (typically broadband)

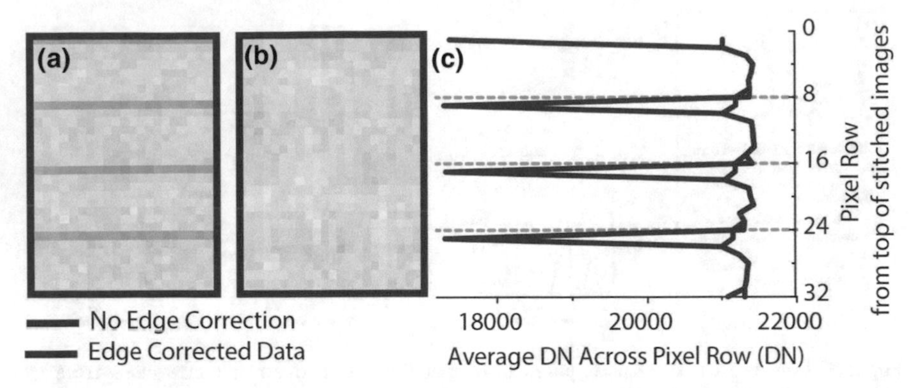

Fig. 3.4 Edge correction was performed to account for the lower QE of pixels rows at the edges of the spectral bands of the Imec linescan sensor. **a** A line artefact is observed in the non-edge corrected spectral band images due to the lower QE of pixel rows at the edges of the spectral bands. A non-edge corrected spectral band image (11) of a reflectance target illuminated with the 660 nm LEDs is here used to illustrate the edge effect. **b** The same image area after edge correction: the line artefact has been removed. **c** The mean DN output across the pixel rows for the non-edge corrected and the corrected spectral band images. For this representative example, the range of pixel values of pixel rows within the same spectral band decreased from 20 to 3% following edge correction

light source by dividing the acquired HSI data by the corresponding images of a reflectance target illuminated in reflectance mode. This approach was however not possible in the current system due to the spectral selectivity of the HSI sensor and the narrow spectral range of the illumination light. LED light can therefore only be used to correct a limited number of spectral bands and will not provide information to correct the fluorescence imaging bands. A broadband uniform liquid crystal display (LCD) screen (White Screen, Xperia-Z3; Sony) was therefore instead placed in the sample plane and imaged and pre-processed by dark subtraction and edge correction. The image of the broadband LCD screen in each spectral band was normalised and inversed to extract a 'vignetting correction' mask for each spectral band (Fig. 3.5). Data were vignetting corrected by multiplying each reconstructed spectral band image with its associated 'vignetting correction' mask.

Reflectance Removal in Software

To account for presence of spectral contributions other than those arising from fluorescence (Fig. 3.6a)—such as the reflectance of the LED excitation light—two software background removal methods were tested and compared to the inclusion of crossed linear polarisers in hardware.

The 'brute force' software approach mimics the action of applying optical bandpass filters to block the reflectance light. The maximum signal intensity of the reflected LED illumination for the power levels chosen in this study was first determined by imaging the uniform reflectance target under standard illumination conditions. Bands where the signal intensity exceeded 10% of the measured maximum reflectance intensity (spectral band 9–17 and 21–35 at F/# 1.65) were then rejected

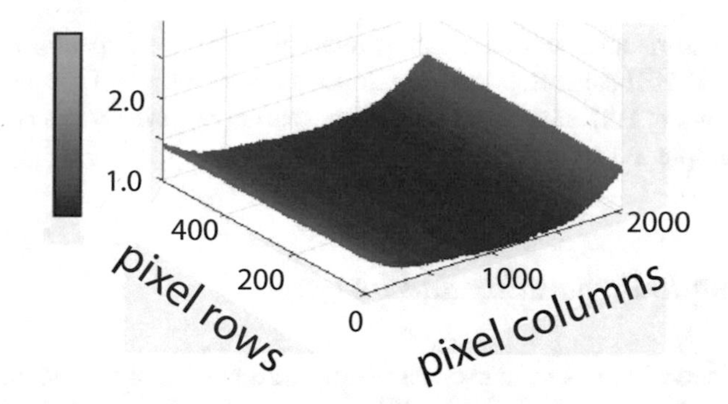

Fig. 3.5 The vignetting correction masks for each spectral band was calculated by imaging a broadband uniform LCD screen. The image show the representative vignetting correction mask for spectral band 11

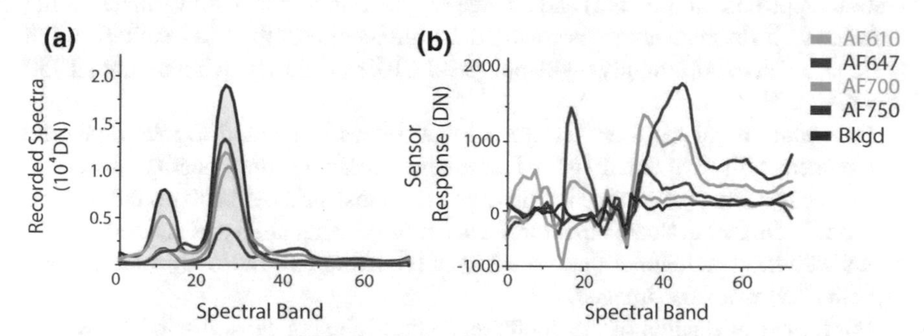

Fig. 3.6 **a** The combined reflectance and fluorescence spectra of the main fluorescent dyes used in this study (AlexaFluor 610, 647, 700 and 750) and a background reflectance spectrum. The acquired fluorescent dye spectra are heavily dominated by the background reflectance signal from the two LED colours within the spectral range of the SRDA. **b** OSP background removal in software is here shown to decouple the background reflectance signal from the fluorescence spectra of the dyes. The fluorescence from the dyes, which were previously dominated by the reflectance from the two LED colours, is now apparent in their reflectance spectra. Successful OSP background removal is also evident from the flat profile of the background spectrum

from the analysis, prior to spectral unmixing. The spectral bands rejected were kept fixed throughout the study.

The 'brute force' reflectance removal method was compared to statistical reflectance removal via OSP (Fig. 3.6b). OSP uses the 'flat' HSI cube to obtain the five most significant principal components of the background. The background principal components were then removed from the spectral endmembers and the acquired HSI according to Eq. 3.3 prior to spectral unmixing.

Spectral Unmixing

Endmembers for spectral unmixing of the fluorescent dyes were extracted from the pre-processed 'reference' HSI data cube by manually placing regions of interests (ROIs) (20 pixels radius) over the known spatial position of the fluorescent dyes. The background endmember was obtained by averaging the pixel spectra from the pre-processed 'flat' HSI cube. Separate dye dilutions were prepared for the acquisition of each 'reference' HSI cube, meaning that endmembers were always collected from a separate sample to that classified.

3.2.3 Sample Preparation and Imaging

Fluorescent dye dilutions in microwell plates were used as a simple and well controlled sample for evaluation of the fHSI system performance. The four main fluorescent dyes used to assess the system performance were AlexaFluor (AF) 610, 647, 700 and 750. These dyes were chosen as they have a relatively high spectral peak separation (approximately 50 nm) and a range of quantum yields, with variable overlap between their fluorescence emissions and the reflectance of the LED excitation light (AF647 and AF700, strongly overlapping; AF610, moderately overlapping; AF750, separated).

Dye dilutions of the four AF dyes in well plates were used to investigate; the best camera objective lens F/# for fluorescence imaging; the linearity of the system response; the system repeatability; and the most suitable background removal method. To further evaluate the fluorescence multiplexing ability of the system, well plates with three additional fluorescent dyes (Cyanine 7.0; Sulfo-Cyanine 7.0; and Cyanine 7.5) were also imaged.

For further evaluation of the fluorescence multiplexing capability of the system, the AF dyes were also imaged in a tissue mimicking agarose phantom. Finally, the four AF dyes were subcutaneously injected in a nude mouse to demonstrate multiplexed fluorescence imaging in vivo.

Fluorescent Dyes in Well Plate

40 μM dilutions of AF610, AF647, AF700 and AF750 (A30050, A20006, A20110 and A37568, all NHS ester; Invitrogen) were prepared in phosphate buffered saline (PBS) (10010015; Thermo Fisher). A well plate (18 well, μ-slide, 81826; ibidi) containing 30 μL of each dye in separate wells, along with PBS controls, was used to acquire the 'reference' HSI cube (with and without crossed linear polarisers). A 'flat' HSI cube of a well plate containing only PBS was recorded to enable OSP background removal. The well plates were imaged at a set of camera objective lens F/#s (F/# 1.65, 2.0, 2.8 and 4.0), to provide endmembers for unmixing the fHSI data acquired at different camera objective lens F/#s.

Dilution series of each dye (1:2 steps from 10 μM to 625 nM) and mixtures of the four dyes were imaged (with and without crossed linear polarisers) to evaluate the linearity of the system response. Duplicates of each dilution series, also containing PBS blanks, were pipetted in opposite directions along two rows of the well plate. Absorption and fluorescence spectra as well as the accuracy of the dilutions were verified using microplate readers (FLUOstar and CLARIOstar; BMG LABTECH). The dilution series were successively imaged at a set of camera objective lens F/#s

(F/# 1.65, 2.0, 2.8 and 4.0) to investigate the impact of F/# on the fluorescence detection efficiency and spectral unmixing performance of the system. Mixtures of the fluorescent dyes—from initial concentrations of 40 μM—were additionally imaged to evaluate the system's ability to delineate the fluorescence signal from spatially co-localised dyes.

To further test the flexibility of the system, a well plate containing three additional fluorescent dyes, Cyanine 7.0; Sulfo-Cyanine 7.0; and Cyanine 7.5 (Cy7, S-Cy7 and Cy7.5, 25090, 25390, 26090; Lumiprobe), was also imaged. A well plate containing 20 μL of 40 μM pure dye dilutions of the AF and cyanine dyes prepared in ethanol (to avoid aggregation of the cyanine dyes) were imaged following the standard imaging (F/# set to 1.65) and pre-processing protocol. A 'reference' data cube with 40 μM pure dye dilutions of the AF and cyanine dyes was also imaged.

To quantify the system repeatability, a well plate containing 4 wells of 30 μL of 40 μM dilutions of AF647 and 750 was imaged 8 subsequent times over a 35 min time period. Well plates containing 30 μL of 40 μM dilutions of AF610, 647, 700 and 750 were also imaged on four separate occasions over a 6 month period and the average coefficient of variation of the unmixed fluorescence signal from the dyes over the different imaging sessions calculated. Data were acquired (F/# set to 1.65) and pre-processed following the standard imaging protocol with OSP reflectance removal.

The performance of the pre-processing and reflectance removal steps were quantitatively compared using well plate fHSI data. ROIs of 65 pixel radius were manually placed over the wells of the well plate to extract an average least squares (LS) score from the abundance maps produced after spectral unmixing. For consistency, the same ROIs were used for all abundance maps. For each dataset areas of strong specular reflections and glare were excluded from the ROIs via a ratiometric thresholding method described later in Sect. 3.2.3. The average LS scores of the duplicate dilutions were used to asses the spectral unmixing performance, with the error in the extracted LS score taken as the range between the duplicates. We also defined the spectral unmixing precision (SUP) to assess the spectral unmixing accuracy to compare the unmixing score assigned to a fluorescent dye endmember to that incorrectly assigned to the other fluorescent dyes. For example, the spectral unmixing precision for AF610 would be the ratio of the signal from an AF610 containing well recorded in the AF610 abundance map and the sum of all signals recorded from the same well in the AF610, AF647, AF700 and AF750 abundance maps. The NNLS score assigned to the background endmember spectrum was not included in the SUP. The background endmember was not included since a LS score assigned to the background spectra does not necessarily indicate a deviation from the ground truth, since strong reflectance signals are present in the fHSI data.

The linearity of the sensor response to the fluorescent dye dilution series was also used to assess the potential for quantitative fluorescence imaging.

Fluorescent Dyes in Tissue Mimicking Phantom

Phantoms were designed to enable further evaluation of the performance of reflectance removal methods on more clinically relevant backgrounds. The phantoms were also used to assess the depth sensitivity of the wide-field fHSI system. For phantom experiments, 1.5% agar solution (05039- 500G; Fluka) was mixed with intralipid emulsion (20% emulsion, I141-100ML; Sigma-Aldrich) to provide a scattering coefficient of $5\,\text{cm}^{-1}$ at 633 nm. A 0.75% volume of $0.5\,\text{mgml}^{-1}$ nigrosin (198285-25G; Sigma-Aldrich) was then added to provide an absorption coefficient of $0.05\,\text{cm}^{-1}$ at 633 nm. While liquid, the phantom base material was poured into multiple petri dishes (664160; Greiner BioOne) to depths of either 0.2 or 1.0 cm. Fluorescent target inclusions were made using transparent plastic straws with an internal diameter of 0.3 cm (391SIPCL; Plastico). The transparent straws were cut into 1 cm long pieces, and one end of each straw piece sealed with a glue gun (PA6-GF30: Type PXP 06; Henkel Pattex Supermatic). Eight 1 cm long straw pieces were inserted into a 1 cm thick phantom slab, with the glue sealed end directed towards the bottom of the petri dish. Half of the straw inclusions were filled (approximately 50 μL) with a 40 μM solution of each of the fluorescent dyes (AF610, AF647, AF700 and AF750) dissolved in equal part phantom base material and PBS. The other half of the straw inclusions were filled with PBS and phantom base material. A second equivalent phantom was prepared with all the straw inclusions filled with the PBS and phantom base material mixture for the 'flat' HSI cube acquisition. The recipe for the phantom base material was developed by Dr. James Joseph (Postdoctoral Researcher, University of Cambridge, UK).

To determine the depth sensitivity of the fHSI system, the 1 cm tissue mimicking slab containing the fluorescent dye inclusions ('base layer') was repeatedly imaged under the fHSI system, as approximately 2 mm thick agarose slabs were placed on top of the 'base layer' (from 0 cm, up to 1 cm in total). The complete imaging protocol was repeated for a slab containing only PBS inclusions. The HSI cube of the 'base layer' of this phantom was taken as the 'flat' background cube in subsequent data analysis.

The fHSI cubes of the slab phantoms were preprocessed and unmixed as per well plates. Endmember spectra for the spectral unmixing of fluorescence signal from the dye inclusions were obtained by imaging a separate well plate with 40 μM dilutions of the AF dyes; the background endmember spectrum was taken as the mean spectra of the 'flat' HSI cube. The ROIs over the fluorescent dye inclusions in the phantom were manually defined by placing 25 pixel radius circular ROIs over the fluorescent inclusion using the HSI cube of the base layer as a reference. The background was defined as the region outside of 50 pixel radii ROIs placed over the inclusions. The definition of the fluorescent inclusion and background ROIs made for the base layer was then propagated to the fHSI data cubes acquired with successive additions of the 2 mm phantom slabs (in which the position of the inclusions was not apparent without spectral unmixing). For each dataset, areas of strong specular reflections and glare were excluded from the ROIs via a ratiometric thresholding method described later in Sect. 3.2.3.

To evaluate the detection sensitivity, the contrast to noise ratio (CNR) of the fluorescent inclusions with depth was calculated. The contrast to noise ratio of each ROI was calculated using the weighted CNR, defined as;

$$CNR = \frac{\mu_{ROI} - \mu_{bkg}}{(w_{ROI}\sigma_{ROI}^2 + w_{bkg}\sigma_{bkg}^2)^{1/2}}, \tag{3.4}$$

taking the signal as the LS score correctly assigned to the fluorescent dye within the ROI (μ_{ROI}) and the background as the LS score incorrectly assigned to the background (μ_{bkg}) in the respective abundance maps. The background subtracted signal is divided by the summed variance of the signal in the ROI (σ_{ROI}) and the background (σ_{bkg}) weighted according to their respective fractional areas (w_{ROI} and w_{bkg}) [47].

Specular Reflectance and Glare Removal in Well Plate and Phantom Data

Ratiometric thresholding was applied to address the presence of specular reflections and glare in well plate and tissue mimicking phantom fHSI data. In these datasets, 1–5% of the image pixels were affected by specular reflections and glare. In the well plate data areas of strong specular reflections and glare probably arose from meniscus effects in the droplets containing the fluorescent dyes. In the tissue mimicking phantom the main source of specular reflections and glare appeared to be moist and wet areas on the phantom surface. In contrast, ratiometric thresholding was not required for the in vivo data since the diffusely reflective mouse skin was not as susceptible to specular reflections and glare (<0.003% of pixels).

The ratiometric thresholding method was developed to remove areas of strong specular reflection and glare from data acquired without the crossed linear polarisers. The specular reflection removal was based on the assumption that specular reflections and glare will skew the intensity ratio of the response within the two spectral bands responding maximally to the 660 and 732 nm LED colours (12 and 27). A ratiometric image was produced by dividing the pre-processed image of spectral band 12 with band 27 (see Fig. 3.7). The mean and standard deviation of the pixels within the ratiometric images were calculated, and pixels outside the qualitatively determined

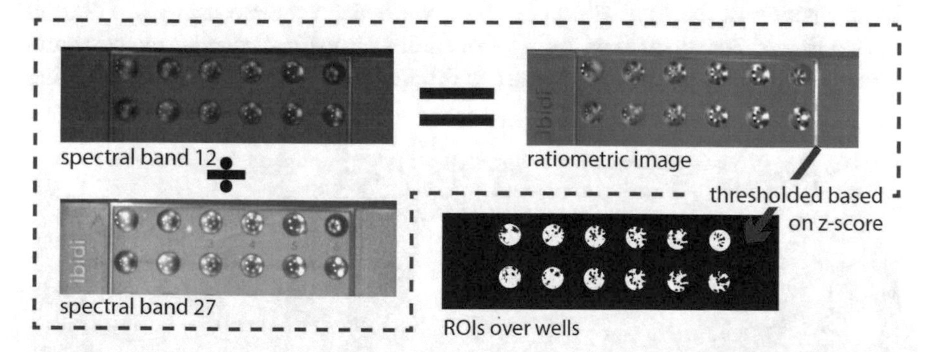

Fig. 3.7 A binary mask was calculated and used to remove the effects of specular reflection and glare from the fHSI data. The binary mask was produced by dividing the pre-processed spectral band image 12 with band 27. The resultant ratiometric image was thresholded by a z-score value of 1.5 to create a binary threshold mask. This figure shows the ROIs over the wells of a well plate after rejection of pixels suffering from specular reflections and glare

z-score value of 1.5 (at which a majority of the specular reflections and glare were visually observed to have been removed) rejected from further analysis. Note that the z-score of a pixel indicates how many standard deviations its value is from the mean pixel value across the full ratiometric image.

Fluorescent Dyes In Vivo

To assess the in vivo imaging performance of the wide-field fHSI system, we administered the four AF fluorescent dyes separately, and in mixture, through subcutaneous injections in nude mice. This allowed us to assess the fluorescence unmixing performance in a real biological background, and to evaluate the potential use of the system as a small animal preclinical imaging platform for in vivo multiplexed fluorescence measurements.

All animal procedures were conducted under a project license issued under the United Kingdom Animals (Scientific Procedures) Act, 1986 (70-8214). The project license was reviewed by the local Animal Welfare and Ethical Review Board and the specific experimental protocol was approved locally by the Named Animal Care and Welfare Officer at the Cancer Research UK Cambridge Institute under compliance form number CFSB0964 and conducted by individuals holding personal licenses. Animal experiments were conducted and designed together with Dr. Isabel Quiros-Gonzalez (Postdoctoral Researcher, University of Cambridge, UK).

40 μM pure dilutions of each of the AF dyes and an equal parts mixture of all four dyes (from 30 μM dilutions) were prepared in a 1:1 solution of PBS and phenol-red-free matrigel (356231; Corning). Experiments were performed in nude mice (n = 2) maintained under inhaled anesthesia using isoflurane mixed with 100% oxygen for a maximum of 2 hours for the entire study. During imaging on the fHSI system, the mouse was placed on a heat pad (Physiosuite Monitoring Module, Kent Scientific Co.) and maintained in a bespoke 3-D printed animal holder (Fig. 3.8). 30 μL of each dye preparation was injected subcutaneously using a Hamilton syringe into the flank of the mouse. The contralateral side of the mouse was used as an imaging control for acquisition of the 'flat' HSI cube. Data were also acquired using an IVIS200 (Perkin Elmer) per standard operating procedures to verify successful injection and placement of the fluorescent dyes using a traditional filter-based fluorescence imaging

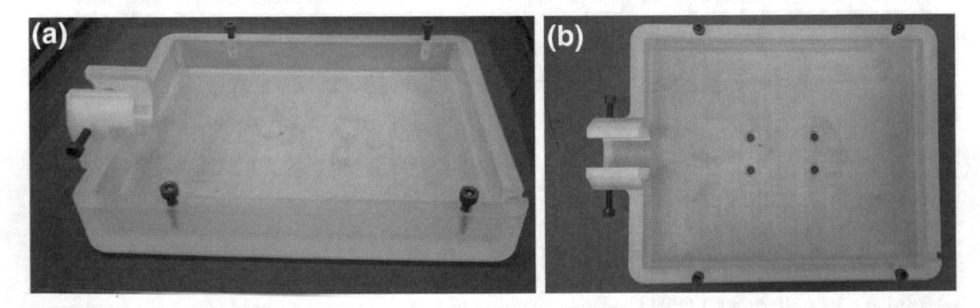

Fig. 3.8 Top (**a**) and side view (**b**) of the bespoke small animal holder for the fHSI in vivo experiments

system. Endmember spectra for the spectral unmixing of fluorescence signals from the dye inclusions were obtained by imaging a separate 'reference' well plate with 40 μM dilutions of the AF dyes; the background endmember spectrum was taken as the mean spectrum of the 'flat' HSI cube.

3.3 Results

3.3.1 System Characterisation

Illumination Uniformity

Quantitative fluorescence imaging across the FOV of the fHSI system is dependent on spatially and temporally uniform sample illumination. To characterise the spatial non-uniformities of the sample illumination provided by the LED ring, the illumination intensities of the LEDs were mapped by translating a calibrated photodetector in 0.5 cm steps across the FOV. Spatial mapping was performed with (Fig. 3.9a) and without (Fig. 3.9b,c), the inclusion of crossed linear polarisers. The spatial non-uniformity did not exceed 5% for the LEDs without crossed linear polarisers. The addition of crossed linear polarisers (and removal of the diffusion paper) substantially affected the uniformity of the illumination, which increased to 29% at 660 nm due to a combination of the removal of the diffusion paper and the low quality of the polarising filters. It is however noted that the sample illumination uniformity with

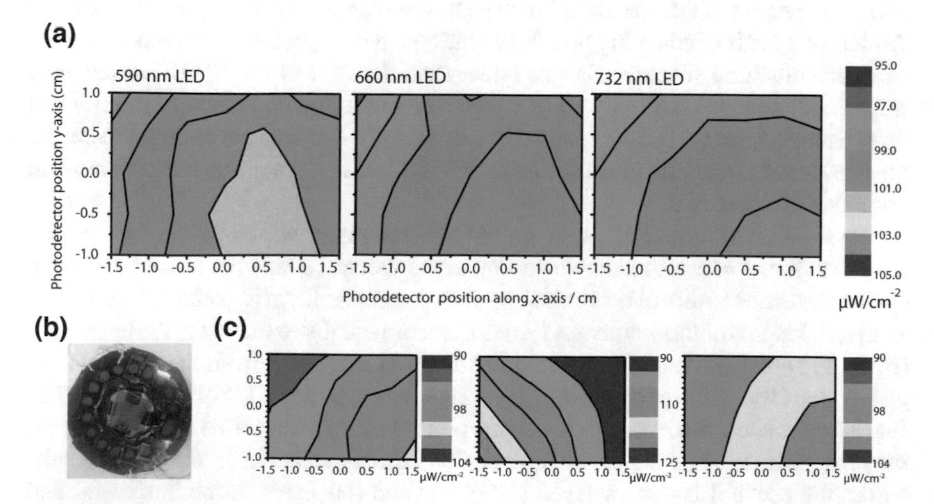

Fig. 3.9 Spatial illumination uniformity of the LED illumination in sample plane. **a** 2D maps of the x–y illumination spatial uniformity in the sample plane. **b–c** The addition of linear polariser film in front of the LED ring adversely affects the spatial illumination uniformity in the sample plane

crossed linear polarisers in place, could be further optimised by using higher quality linear polarisers and by improving the mounting and alignment of the polarisers.

The intensity drift of the LEDs over a 15 min time period was less than 2%. The temporal intensity drift of the LEDs was thus considered insignificant in comparison to the intensity differences due to spatial non-uniformities of the sample illumination.

Keystone and Smile

Traditional linescan HSI sensors based on diffraction gratings frequently suffer from aberrations, such as spectral smile and keystone [44]. Although we expected such aberrations to be minimal since the spectral filters were here monolithically integrated on sensor, these aberrations were characterised for experimental completeness. Spectral smile refers to a shift in the peak wavelength of the spectral response of pixels in the same row. This aberration was characterised by imaging (F/# set to 1.65) a uniform reflectance target illuminated with 660 nm LED light and extracting the spectral response across the SRDA.

Mean ROI spectra were extracted from 64 ROIs of 8 pixel width equally spaced across the acquired spectral band image in a direction orthogonal to the scan direction. As a uniformly reflective target was imaged, the ROIs were extended across the whole FOV in a direction parallel to the scan direction. Assuming uniform reflectance from the target, this served as taking a temporal average over the acquisition time.

The spectral band which responded maximally to the reflectance light was extracted from each ROI to investigate the potential presence of a spectral smile aberration (Fig. 3.10). We noted that the reflectance peak of the ROIs across the FOV differed by a maximum of 2 spectral bands; the reflectance spectra either peaked in band 11 (peak $\lambda \pm$ FWHM $= 659 \pm 18.8$ nm), 12 (653 ± 18.4 nm) or 13 (653 ± 18.1 nm). It was further noted that the variation in peak wavelength was not symmetric around the sensor's centre, indicating that the variations in the spectral band response arose from manufactural sensor variances rather than the AOI of the light. The detected wavelength change ($\Delta\lambda = 6$ nm) was however small in comparison to the FWHM (approximately 18 nm) of the spectral band response. Spectral smile could therefore be considered negligible given the broad FWHM of the spectral band responses in our mode of operation.

Keystone aberrations appears as spatial shifts of pixels with the same peak wavelength response, causing spatial distortion of the spectral band images [44]. Potential spatial distortions were characterised by imaging a regular grid pattern (Fig. 3.11a). The pixel lengths of the regular grid pattern were manually extracted in the horizontal (Fig. 3.11b) and vertical (Fig. 3.11c) directions, as measured from the centre of the grid pattern for a subset of spectral band images (1, 9, 18, 27, 36, 45, 54 and 72). Fig. 3.11b, c show the mean pixel lengths plotted against the actual distance in mm, the error was taken as the standard deviations of the variation in the pixel lengths across the spectral bands analysed (1.6 pixels and 2.0 pixels in the horizontal and vertical direction respectively; error bars contained within the points of the scatter plots). The high R^2 values and low RMSE of both linear fits ($R^2 = 1$; 2.69 and 1.89 pixels) indicated the absence of keystone aberrations in both the horizontal and vertical directions.

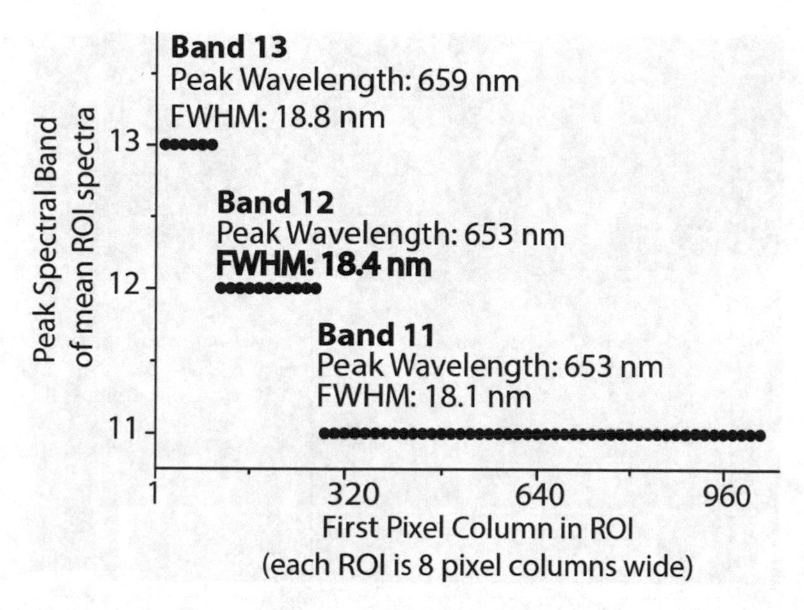

Fig. 3.10 Spectral smile was investigated by imaging a reflectance target illuminated with 660 nm LEDs. The peak wavelength and FWHM of ROI spectra, extracted from 8 pixel wide ROIs placed across the FOV, show that the spectral band responding maximally to the reflected light changes from band 13 to 11; this corresponds to a wavelength change of 6 nm. Given that the FWHM of the band response was approximately 18 nm, the spectral smile was considered negligible

3.3.2 Well Plate Data

fHSI at Different Camera Objective Lens F/#s

No clear trends could be observed in performance of the LS scores and spectral unmixing precision when changing the objective lens F/# between 1.65 and 4.0 (Fig. 3.12). However, as shown in the sensor calibration (Sect. 2.3.3), increasing the F/# limits the optical throughout of the system. For AF610, the combined optical throughput and sensitivity were indeed too low to detect the dye emissions at a F/# of 4.0. For fluorescence imaging, with low light levels and broad spectral profiles, we thus selected a low F/# to maximise the optical throughput. In the remainder of this work data acquired at an F/# of 1.65 are presented.

Reflectance Removal Methods

We evaluated the impact of different background removal methods used in image pre-processing on the data quantification. Figure 3.13a shows the LS score recorded for the four fluorescent AF dyes at 10 μM; endmembers at 40 μM concentration were input to the NNLS so a LS score of 0.25 would therefore be expected at 10 μM, with 0 for the background. In all cases, applying NNLS without background removal resulted in an incorrect LS score assignment to both the signal and background, with the exception of the signal from AF750, for which the fluorescence signal was well

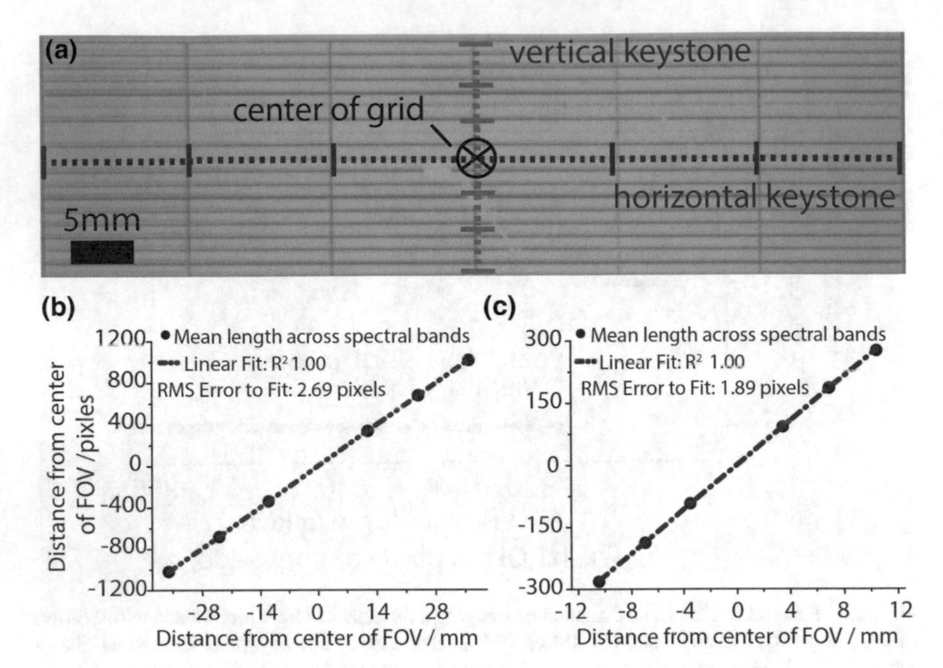

Fig. 3.11 A regular grid pattern was imaged to investigate the potential presence of keystone aberrations in the wide-field fHSI system. **a** The mean pixel lengths for a subset of spectral band images (1, 9, 18, 27, 36, 45, 54, and 72; spectral band 27 shown in figure) were plotted against the known dimension of the grid pattern in the horizontal (**b**) and vertical (**c**) directions. The high linearity ($R^2 = 1$) and low RMSE (2.69 and 1.89 pixels) of the fits indicate that no keystone aberrations were present in either the vertical or horizontal directions

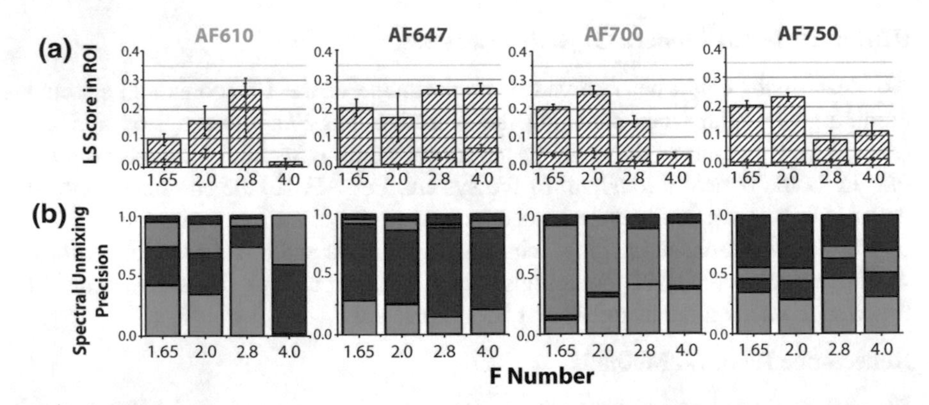

Fig. 3.12 **a** The LS score assigned to the ROIs placed over wells of a well plate (black, dye; red, background = PBS) imaged with the fHSI system for different objective lens F/#s. For AF610, the combined optical throughput and sensitivity is too low to detect the dye emissions at a F/# of 4.0. The error bars indicate the range of the LS scores measured from duplicate dye dilutions and PBS wells in the same well plate. **b** The spectral unmixing precision of the dyes for increasing F/#s. The figure shows data from 10 μM dilutions of the fluorescent dyes, unmixed using endmembers extracted from 40 μM dye dilutions. Data were pre-processed using OSP

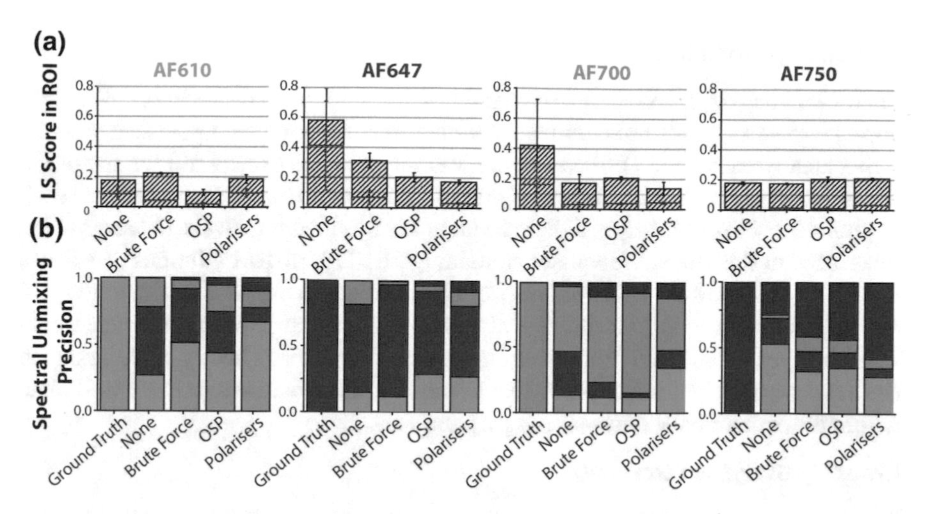

Fig. 3.13 fHSI accurately identifies each fluorescent dye and is able to perform multiplexing. **a** NNLS spectral unmixing was applied to well plates containing pure solutions of the different fluorescent dyes at 10 μM concentration, with and without background removal. The LS score from the dye abundance map within the expected well (black striped bars) and in a background well containing PBS (red bars) are compared. The error bars indicate the range over two dye and PBS wells in the same well plate. **b** Spectral unmixing precision (SUP) is the ratio of the LS score for the correct abundance map to the sum of scores recorded from the same well across all other fluorescent dye abundance maps (but not the background abundance map). For example, in the case of a well containing AF610, the LS scores are used to calculate the SUP as follows: SUP = AF610 / (AF610 + AF647 + AF700 + AF750). The ground truth for each pure dye dilutions gives SUP = 1

separated from the reflectance light. AF647 and AF700 show poorest performance without background removal, which may be explained by the strong overlap of the AF647 and AF700 fluorescence emission with the reflectance light. All background removal methods improved this to some extent, however, 'brute force' removal and crossed linear polarisers in hardware both had high misfitting to the background. The statistical method of OSP provided the optimal combination of accuracy in the signal and background, although the LS score was consistently underestimated. In all cases, the AF610 dye was least well resolved, most likely due to the low detection sensitivity of the HSI camera in this spectral region and strong spectral overlap with the reflectance light.

In addition to the LS score, we defined the SUP: the ratio of the LS score for the correct abundance map to the sum of scores recorded from the same well across all other fluorescent dye abundance maps. For each background removal method, the SUP of the pure dye dilutions were compared to the ground truth (Fig. 3.13b). Again, the OSP method performed well across the board, qualitatively allowing a majority decision to be made regarding the dye present in a given location in the well plate.

System Repeatability

For repeat imaging of a well plate containing 40 μM dilutions of AF fluorescent dyes during a 35 min period, the coefficient of variation of the unmixed fluorescence signal (reflectance removal via OSP) with time was found not to exceed 6% for any of the wells. Based on the low coefficient of variation, we could conclude that effects such as photobleaching of the dyes did not occur at the illumination levels and acquisition times used in this study. Well plates containing 30 μL of 40 μM dilutions of AF610, AF647, AF700 and AF750 were also imaged on four separate occasions over a 6 month period; the average coefficient of variation of the unmixed fluorescence signal from the dyes over the different imaging sessions was 16%. The larger variations in the fluorescence signals was likely due to the different positioning of the well plate within the camera FOV between imaging sessions.

Linearity of System Response

The linearity of the fHSI response of the system was evaluated by imaging 1:2 dilution series of the AF dyes from 10 μM to 625 nm. It can be concluded that a linear fluorescence response was observed, based on the high R^2 values and low RMSE for all linear fits to the LS scores of the whole dilution series across the range of dyes (Fig. 3.14). As previously, endmembers were defined for 40 μM dye dilutions and put into the NNLS so that a LS score of 0.25 would be expected for thc 10 μM dilutions, with 0 for the background. As previously observed, a LS score of close to 0.25 was retrieved for all the AF dilution series except for the AF610 dilution series. This further indicates that the system has the potential to extract quantitative fluorescence data provided that all dyes can be detected with sufficient SBR. Slight deviations from the expected LS value of 0.25 may have arisen from systematic effects, such as sample illumination non-uniformities.

fHSI Spectral Unmixing Capability

A fundamental limitation of fluorescent dye mixtures with broad overlapping emissions (not limited to our application) is 'bleed over'. This depends on the fact that the emission of one fluorescent dye can immediately excite another in the same solution. For example, the area of overlap between the AF610 emission spectrum and the excitation spectra of the other dyes used may be up to 65, 48 and 28% for AF647, AF700 and AF750 respectively. Thus, while the primary constituents in each well could be identified in most cases (Fig. 3.15), extraction of quantitative concentration data was difficult. AF610 was consistently underestimated in the mixtures, likely due to a combination of low camera sensitivity, strong overlap between AF610's emission spectra and the reflectance light and high fluorescence emission overlap with the absorption spectra of the other dyes.

To assess the response of the system in the case of more overlapping spectra, we introduced a well plate containing 3 additional fluorescent dyes (Cy7, S-Cy7 and Cy7.5) with significant spectral overlap with both the reflectance light and the existing AF dyes. The abundance maps for each fluorescent dye is shown in Fig. 3.16a, b–h. Based on the measured endmembers (Fig. 3.16i), all 7 dyes could be clearly resolved

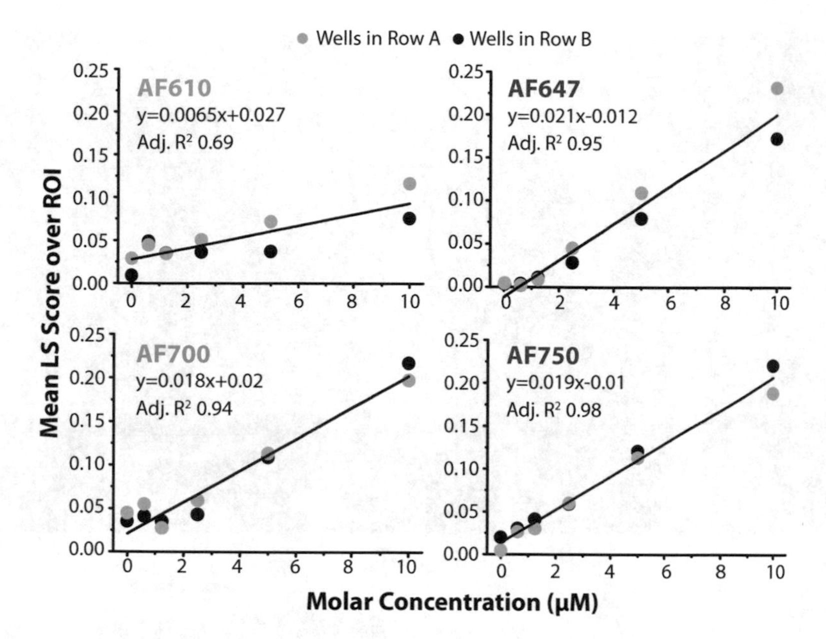

Fig. 3.14 Linearity of response to a 1:2 dilution series from 10 μM of each dye, with associated linear fit parameters. The LS score was expected to be 0.25 at the highest dilution. Data were pre-processed using OSP

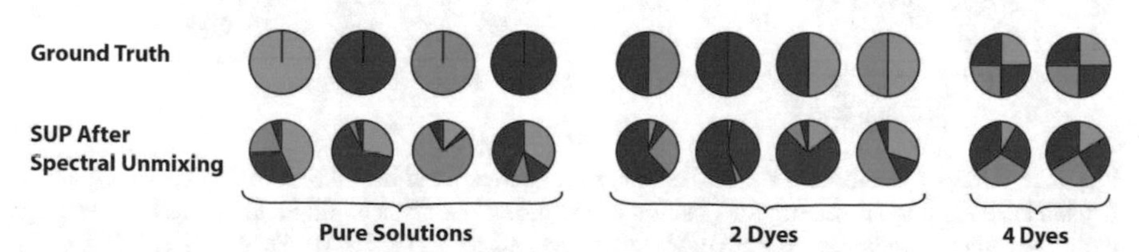

Fig. 3.15 The multiplexing capability of the fHSI system is demonstrated by the SUP of pure dye (average over two wells of 10 μM from dilution series) as well as 2 and 4 component mixtures imaged in a well plate. While the primary constituents of each component mixture could often be determined via majority decision, extraction of quantitative concentration from dye mixtures was difficult. Data have been background removed via OSP

(Fig. 3.16j). A limited degree of misfitting between AF750, Cy7, S-Cy7 and Cy7.5 was observed, which would be expected based on the similarity of their emission spectra.

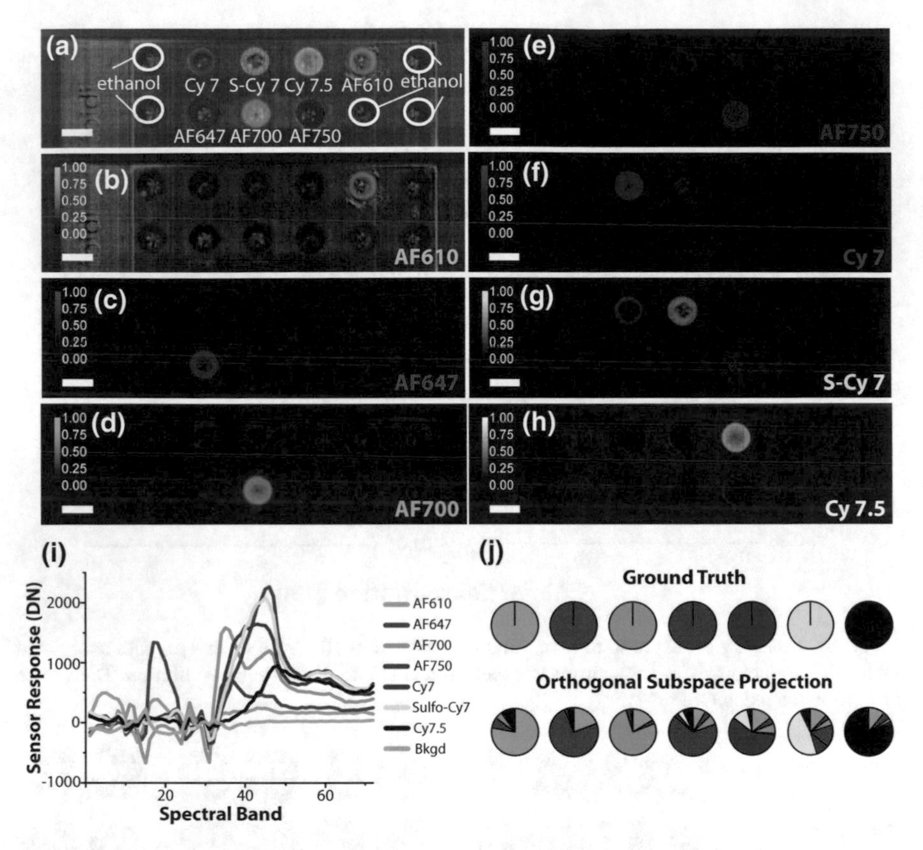

Fig. 3.16 Up to 7 fluorescent contrast agents could be resolved in solution in a microwell plate. **a** A pseudocolour map, revealing the position of the individual dyes could be generated by merging their respective abundance maps: 40 μM concentrations of **b** AF610, **c** 647, **d** 700, **e** 750, **f** Cy7, **g** S-Cy7 and **h** Cy7.5 in ethanol were imaged and unmixed. **i** Endmembers for spectral unmixing of the data. **j** The spectral unmixing precision shows that each dye could be identified via majority decision after spectral unmixing. Spectral data and endmembers were pre-processed using OSP. This figure has been previously published in: A.S. Luthman, S. Dumitru, I. Quiros-Gonzalez, J. Joseph, S.E. Bohndiek, J. Biophotonics 2017, 10, 840

3.3.3 Phantom Data

Tissue mimicking agarose phantoms were used to explore the spectral unmixing performance and depth sensitivity on a more realistic 'tissue' background. The four AF fluorescent dyes were individually placed in an intralipid phantom at a concentration of 40 μM, confined in the phantom using transparent plastic straws. Initially the phantom was imaged with the top of the straw exposed, then agarose gel slabs of up to 2.5 mm thickness were sequentially placed on top to of the base layer to gradually increase the depth of the fluorescent dye inclusions (Fig. 3.17a, b). ROIs of 50 pixel radii were manually placed over the fluorescent dye inclusions to extract the average

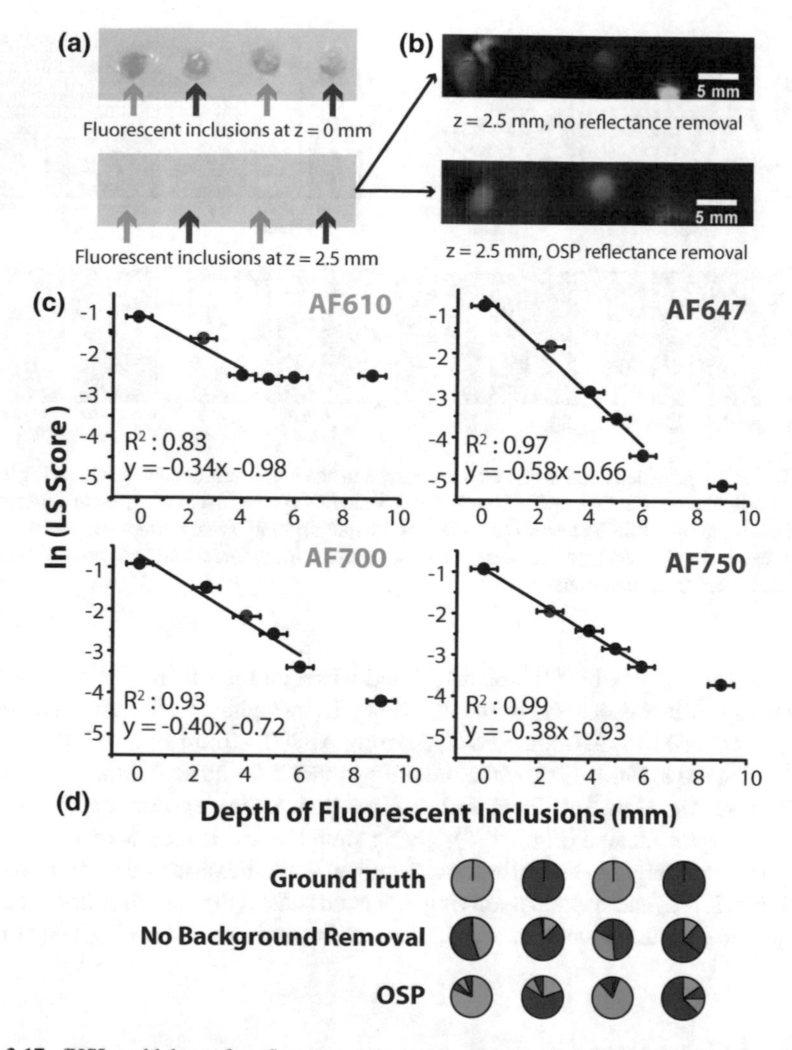

Fig. 3.17 fHSI could detect four fluorescent dyes at depth in tissue mimicking phantoms. **a** Photograph of the phantom containing AF dye inclusions before and after the addition of a 2.5 mm thick slab of tissue mimicking agarose gel. **b** Pseudocolour abundance maps for each dye with and without OSP background removal in pre-processing. The maximum intensity value in each fluorescence abundance map was rescaled to a LS score of 1 for visualisation. **c** The LS score of the fluorescent inclusions as a function of depth. The red data markers indicate the last data point with a weighted CNR of 4 or higher. Error bars indicate the measurement uncertainty of the thickness of the agarose gel slabs. **d** A comparison of the spectral unmixing precisions show that whilst it was possible to identify the dyes based on a majority decision for the OSP pre-processed data, this was not possible without background removal. SUP data from fluorescent dye inclusions at $z = 2.5$ mm are shown. This figure has been previously published in: A.S. Luthman, S. Dumitru, I. Quiros-Gonzalez, J. Joseph, S.E. Bohndiek, J. Biophotonics 2017, 10, 840

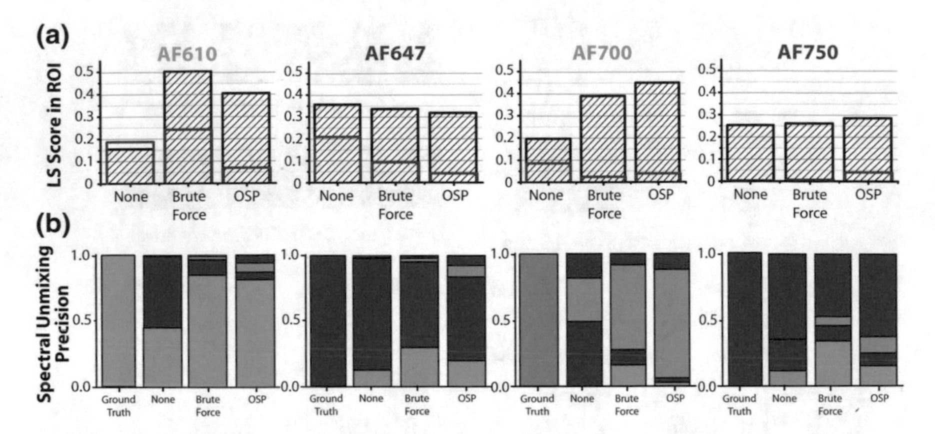

Fig. 3.18 **a** OSP provided optimal reflectance removal in phantoms by reducing the misfitting to the background without impacting the SUP. **a** LS score within ROIs placed over the dye inclusions (black striped bars) and over PBS inclusions (red bars) at 2.5 mm depth in tissue mimicking phantom. LS score is shown for no reflectance removal (none), as well as 'brute force' and OSP pre-processing. **b** SUP compared to ground truth

LS score. The weighted CNR was calculated based on the LS from ROIs over the fluorescent dye inclusions and the background. The weighted CNR remained above 4 to depths of: AF610, 2.5 mm; AF647, 2.5 mm; AF700 4.0 mm and AF750, 2.5 mm (indicated in red in Fig. 3.17c), while linearity persisted for up to a further 4 mm with high R^2 values and low RMSEs of the linear fits (Fig. 3.17c). Spectral unmixing precision was also evaluated (Fig. 3.17d), giving similar performance as in well plates. The benefit of applying OSP reflectance removal could be observed in both the LS scores (Fig. 3.18a) and spectral unmixing precision values (Fig. 3.18b) extracted from ROIs placed over the fluorescent dye inclusions in the tissue mimicking phantom.

3.3.4 In Vivo Data

To assess the in vivo imaging performance of the wide-field fHSI system, we administered the four AF fluorescent dyes separately, and in mixtures, through subcutaneous injections in nude mice. This allowed us to explore the spectral unmixing performance of the fHSI system on a real tissue background and to evaluate the potential of the system as a small animal preclinical imaging platform.

The locations of all four fluorescent dyes could be retrieved from their respective fHSI abundance maps (Fig. 3.19a). In mixtures (Fig. 3.19b), only three dyes can be clearly observed in the abundance maps, due to the challenges noted above for the detection of AF610 in mixtures. The successful detection of the pure fluorescent dye injections with the fHSI system was reinforced by LS scores and SUP extracted from 30 pixel radii ROIs placed over the injection sites of the separate subcutaneous

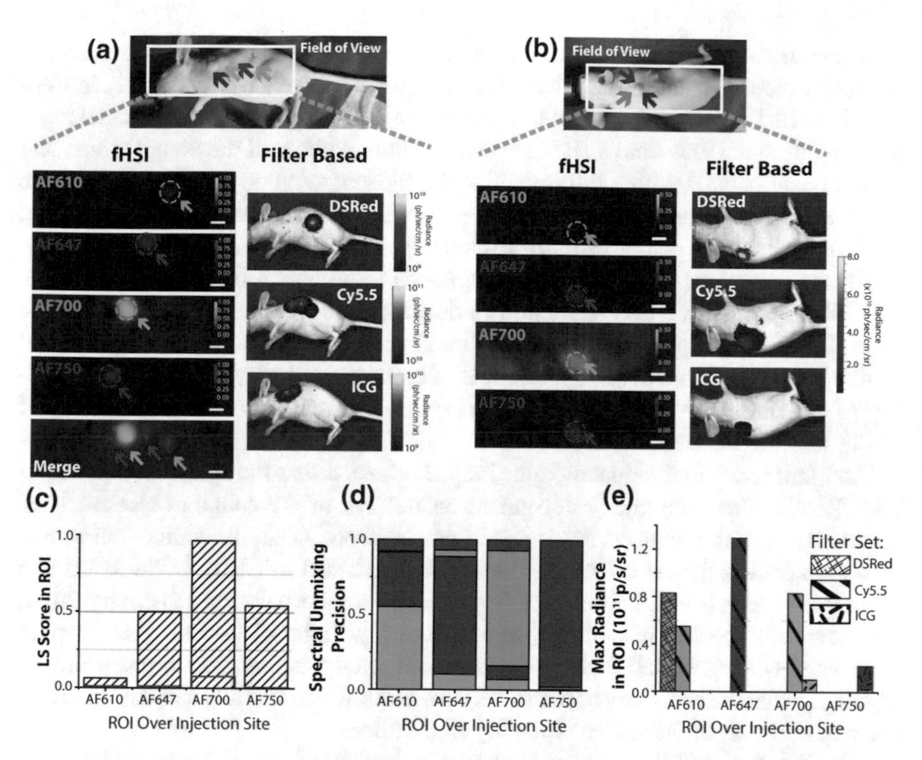

Fig. 3.19 fHSI enables spectral resolution of four AF fluorescent dyes in living subjects. **a** Abundance maps and the pseudocolour image show clear resolution of separate injection sites (indicated by arrows) of each fluorescent dye for the fHSI. **b** Abundance maps from a mixed injection of all four dyes. Intensity scale bar indicates the LS score in both cases with spatial scale bar = 5 mm. Data acquired using a filter based approach (IVIS 200; Perkin Elmer) are shown for reference. ROIs of 30 pixel radii were then placed over the separate subcutaneous injection site on the fHSI data to extract average LS scores (**c**) (background in red defined as all pixels outside of the injection site ROIs), spectral unmixing precision (**d**) and from the filter-based data, maximum radiance (**e**). This figure has been previously published in: A.S. Luthman, S. Dumitru, I. Quiros-Gonzalez, J. Joseph, S.E. Bohndiek, J. Biophotonics 2017, 10, 840

dye injections (Fig. 3.19c, d). As would be expected, the equivalent quantification from the filter-based optical imaging instrument (IVIS 200; Perkin Elmer) showed significant overlap between the recorded fluorescent dye emissions (Fig. 3.19e).

3.4 Discussion and Conclusions

This chapters presents the design, characterisation and imaging performance of a fHSI system based around a SRDA linescan sensor. The system has intrinsic potential as a fHSI imaging platform, as it may be used for multiplexed fluorescence imaging of a wide range of fluorescent samples. With further developments the system could, for example, be used as a small animal imaging platform for preclinical research, or to rapidly measure multiplexed fluorescence signals from excised tissues. Additionally, the fHSI system allowed us to explore the software and hardware challenges of integrating a SRDA into a fHSI system, without adding all the complexities that would be associated with a fully functional endoscopic/intraoperative system. The reflectance based illumination geometry—mimicking that typically encountered in clinical applications—allowed us to develop data acquisition and pre-processing protocols for the future straightforward integration into clinical imaging applications. The fHSI system may also function as a development platform for the further use of SRDAs for multiplexed fluorescence imaging. In addition to the work presented in this thesis, the system has, for example, also been used to test an algorithm aiming to determine the optimum number of spectral bands required for unmixing of multiplexed fluorescence data [48].

The sensor calibration detailed in Chap. 2, showed that the spectral response of the SRDA's filters are highly dependent on the F/# of the camera objective lens. In this chapter, the fluorescence imaging performance of the Imec linescan sensor was evaluated at the set of F/# previously characterised in Chap. 2. We found that an objective lens F/# of 1.65 offered a good compromise between sensitivity (high) and adequate spectral resolution (low) for the low light levels and broad spectra associated with the small molecule fluorescent dyes used here. This result further highlights that the accessory optics of SRDAs needs to be carefully selected to match the requirements of the specific imaging application.

We also explored three reflectance removal methods suited for the reflectance based illumination geometry frequently encountered in the clinic; two software methods, 'brute force' removal of the bands encompassing a majority of the reflectance signal, and statistical background removal via OSP; and a hardware method based on cross-polarisers. It was found that reflectance removal via OSP yielded the most accurate retrieval of fluorescence signals in combination with strong suppression of spectral misfitting to the background. In comparison to the removal of specular reflections and glare in hardware via cross-polarisers, software methods also yields 50 % higher light throughput as they do not require polarised light. As a statistical background removal method, OSP also has the advantage of accounting for any systematic noise present in the imaging system. For strongly AF samples, OSP may also allow separation of the intrinsic and extrinsic fluorescence signals, although this was not evaluated in our study. Clinical contrast could then be simultaneously extracted from extrinsically administered fluorescent contrast agents and from the intrinsic tissue AF. Whereas our study shows that OSP reflectance removal has high potential as a pre-processing method for clinically acquired fHSI data, further work is required. The spectral unmixing performance presented here should for example be verified on highly AF samples. There would also be merit in wider explorations of alternative spectral unmixing algorithms to further optimise the fHSI performance.

When applying OSP background removal we demonstrated simultaneous imaging and unmixing of up to seven commonly used fluorescent dyes with highly overlapping emissions in pure solutions. When prepared in mixtures, fluorescent dyes naturally 'bleed over' i.e. the emission of one dye excites another, which affects any attempt to quantify concentration of highly overlapping dyes (independent of the imaging system used). We therefore focused on extracting quantitative data from four relatively well spectrally separated dyes. The presence of individual dyes at a given location could be identified based on majority decisions in our data; in superficial well plates, at depths of several millimetres in tissue mimicking phantoms, and in vivo in mice. We thus demonstrated the performance of our system on a wide range of samples with up to seven fluorescent dyes. Ultimately the system could be easily optimised for the detection of alternative dyes due to the plug-and-play flexibility of the sample illumination allowing arbitrary LEDs to be included in the illumination ring.

Furthermore, we compared our in vivo data to images from a traditional filter-based fluorescence system, which reinforced the advantage of increased spectral sampling for multiplexed fluorescence imaging. It is however difficult to quantitatively compare the imaging performance of different fluorescence systems since there is currently no agreement which standards should be used to assess their performance [35]. This is particularly problematic for characterising the systems for widespread imaging, and naturally when seeking regulatory approval for novel fluorescence imaging system. A more widespread use of phantoms and accepted standards to assess the performance of novel fluorescent dyes and imaging systems are therefore required [35]. Lacking commonly accepted fluorescent standards, we instead sought to develop a flexible imaging system which we evaluated on a wide range of fluorescent samples.

Although our fHSI system shows promise, the system does however have several limitations. Firstly, the system performance is limited by its low fluorescence detection sensitivity. This is partly due to the low QE and dynamic range (DR; the range of light intensities accurately recorded by the camera) of the commercial CMOS image sensor used in the HSI camera. Future integration of the optical filters onto a scientific grade sensor could overcome the QE limitations. The DR could be extended by using a sensor with spatially programmable exposure times or by an application specific filter deposition, designed to attenuate the reflected excitation light. An increased DR could also allow the use of higher sample illumination intensities.

The sample illumination intensity is currently limited by the camera saturation rather than the fluorescence plateau of the dyes or the in vivo maximum permissible exposure illumination limits [49]; higher sample illumination intensities could therefore increase the available fluorescence signal. To achieve higher illumination intensities, the illumination scheme may however need to be modified. Based on ray trace modelling of the current illumination scheme, it has been shown that the illumination efficiency of the LED ring is limited [50]. Should the DR of the camera be increased, the LED ring may still not provide sufficient illumination power to maximise the detection capabilities of the fHSI system. The development of more efficient—and/or higher intensity—illumination schemes for wide field fHSI imaging is therefore an important area for further work. One of the main benefits of SRDAs is the potential for very low manufacturing costs [51]. To maintain this benefit, additional system components would also need to be selected from a cost perspective. It would therefore be beneficial if future illumination schemes could maintain the low cost of the current LED ring solution.

Furthermore, while we were able to demonstrate quantitative imaging in solutions in vitro, achieving this in vivo is more challenging as sample topology and the optical properties of tissue can modulate the recorded fluorescence spectra. The absorption and fluorescence of intrinsic tissue chromophores can affect the magnitude and the spectral profile of the recorded fluorescence signals from extrinsic fluorescent contrast agents, making quantitative spectral unmixing challenging [52, 53]. This "spectral colouring" is particularly problematic when imaging at depth, as the illumination and fluorescence signal propagates further through tissue [54]. Due to the often unknown spectral contribution from intrinsic chromophores, non-linear inversion spectral unmixing algorithms and light fluence correction models are therefore required for quantitative fluorescence imaging in vivo at depth [54]. To achieve this, the imaging protocol and analysis methods of the current fHSI system need to be further developed. Additionally, sample topology should also be considered in order to achieve quantitative fluorescence imaging of extended three dimensional objects; this may include methods to ensure even illumination of three dimensional extended objects [50], corrections for variable working distances [55] and modelling of light propagation in tissue [52–54].

In this work, all data processing was also performed off-line, whilst it for many in vivo imaging applications would be critical to produce the abundance maps in real-time. Transferring our data processing algorithms onto a field programmable array (FPGA) or graphical processing unit (GPU) could increase the processing speeds to allow visualisation of fluorescence abundance maps directly after the spatial scanning. Finally, while pushbroom imaging is acceptable for static (or low temporal resolution) samples (e.g. for optical inspection of excised tissues after surgery), for video rate applications such as endoscopic and intraoperative imaging, it is likely that a pixel level deposition of filters will be preferable, requiring a trade-off between spectral and spatial resolution. In the next chapter of this thesis we therefore move to working with snapshot SRDAs, allowing us to perform endoscopic imaging.

To conclude this chapter, we have integrated a linescan SRDA sensor in a wide-field fHSI system to demonstrate multiplexed fluorescence imaging of seven fluorescent dyes in vitro and at least four dyes in tissue mimicking phantoms and in vivo in a mouse model. Whilst the wide-field system has intrinsic value as a fHSI imaging and development platform, it also allowed us to explore the future implementation challenges of integrating SRDAs in clinical instrumentation.

References

1. R.R. Zhang et al., Beyond the margins: real-time detection of cancer using targeted fluorophores. Nat. Rev. Clin. Oncol. **14**, 347–364 (2017)
2. A.L. Vahrmeijer, M. Hutteman, J.R. van de Vorst, C.J.H. van de Velde, J.V. Frangioni, Image-guided cancer surgery using near-infrared fluorescence. Nat. Rev. Clin. Oncol. **10**, 507–518 (2013)
3. M.B. Strum, T.D. Wang, Emerging optical methods for surveillance of Barrett's oesophagus. Gut **64**, 1816–1823 (2015)

4. J. Hoon Lee, T.D. Wang, Molecular endoscopy for targeted imaging in the digestive tract. Lancet. Gastroenterol. Hepatol. **1**(2), 147–155 (2016)
5. A.S. Luthman, S. Dumitru, I. Quiros-Gonzalez, J. Joseph, S.E. Bohndiek, Fluorescence hyperspectral imaging (fHSI) using a spectrally resolved detector array. J. Biophotonics **10**(6–7), 840–853 (2017)
6. G. Lu, B. Fei, Medical hyperspectral imaging: a review. J. Biomed. Opt. **19**(1), 010901 (2014)
7. S. Keereweer et al., Optical image-guided cancer surgery: challenges and limitations. Clin. Cancer Res. **19**(14), 3745–3754 (2013)
8. M.S. Chin et al., Hyperspectral imaging for early detection of oxygenation and perfusion changes in irradiated skin. J. Biomed. Opt. **17**(2), 026010 (2012)
9. N.T. Clancy et al., Intraoperative measurement of bowel oxygen saturation using a multispectral imaging laparoscope. Biomed. Opt. Express **6**(10), 4179–4190 (2015)
10. F. Vaesfi et al., Separating melanin from hemodynamics in nevi using multimodal hyperspectral dermoscopy and spatial frequency domain spectroscopy. J. Biomed. Opt. **21**(11), 114001 (2016)
11. M.A. Calin et al., Hyperspectral imaging-based wound analysis using mixture-tuned matched filtering classification method. J. Biomed. Opt. **20**(4), 046004 (2015)
12. W. Ren, Q. Gan, Q. Wu, S. Zhang, R. Xu, Quasi-simultaneous multimodal imaging of cutaneous tissue oxygenation and perfusion. J. Biomed. Opt. **20**(12), 121307 (2015)
13. W.R. Johnson, D.W. Wilson, W. Fink, M. Humayun, G. Bearman, Snapshot hyperspectral imaging in opthalmology. J. Biomed. Opt. **12**(1), 014036 (2014)
14. H. Li et al., Snapshot hyperspectral retinal imaging using compact spectral resolving detector array. J. Biophotonics **10**(6–7), 830–839 (2016)
15. L.E. MacKenzie, T.R. Choudhary, A.I. McNaught, A.R. Harvey, In vivo oxiometry of human bulbar conjunctival and episcleral microvasculature using snapshot multispectral imaging. Exp. Eye Res. **149**, 48–58 (2016)
16. T.H. Tate et al., Multispectral fluorescence imaging of human ovarian and fallopian tube tissue for early-stage caner detection. J. Biomed. Opt. **21**(5), 014036 (2016)
17. G. Lu et al., Spectral-spatial classification for noninvasive cancer detection using hyperspectral imaging. J. Biomed. Opt. **19**(10), 016004 (2014)
18. Z. Han, A. Zhang, X. Wang, M.D. Wang, T. Xie, In vivo use of hyperspectral imaging to develop a noncontact endoscopic diagnosis support system for malignant colorectal tumors. J. Biomed. Opt. **21**(1), 016001 (2016)
19. S.V. Panasyuk et al., Medical hyperspectral imaging to facilitate residual tumor identification during surgery. Cancer Biol. Ther. **6**(3), 439–446 (2007)
20. J. Pichette et al., Intraoperative video-rate hemodynamic response assessment in human cortex using snapshot hyperspectral optical imaging. Neurophotonics **3**(4), 045003 (2016)
21. Q. Li et al., Review of spectral imaging technology in biomedical engineering: achievements and challenges. J. Biomed. Opt. **18**(10), 100901 (2013)
22. G. Hong, A.L. Antaris, H. Dai, Near-infrared fluorophores for biomedical imaging. Nat. Biomed. Eng. **1**(0010), 1–22 (2017)
23. S.L. Jacques, Optical properties of biological tissues: a review. Phys. Med. Biol. **58**, 5007–5008 (2013)
24. F. Leblond, S.C. Davis, P.A. Valdés, B.W. Pouge, Pre-clinical whole-body fluorescence imaging: review of instruments, methods and applications. J. Photochem. Photobiol. B **98**, 77–94 (2009)
25. S.H. Yun, S.J.J. Kwok, Light in diagnosis, therapy and surgery. Nat. Biomed. Eng. **1**(0008) (2017)
26. H. Hyun et al., Structure-inherent targeting of near-infrared fluorophores for parathyroid and thyroid gland imaging hoon. Nat. Med. **21**(2), 192–197 (2015)
27. J.P. Miller, D. Maji, J. Lam, B.J. Tromberg, S. Achilfu, Noninvasive depth estimation using tissue optical properties and a dual-wavelength fluorescent molecular probe in vivo. Biomed. Opt. Express **8**(6), 3095–3109 (2017)

28. M. Lakadamyali, Super-resolution microscopy: going live and going fast. ChemPhysChem **15**(4), 630–636 (2014)
29. T. Barrett et al., In vivo diagnosis of epidermal growth factor receptor expression using molecular imaging with a cocktail of optically labeled monoclonal antibodies. Clin. Cancer Res. **13**(22), 6639–6648 (2007)
30. J.R. Mansfield, Multispectral imaging: a review of its technical aspects ad applications in anatomic pathology. Vet. Pathol. **51**(1), 185–210 (2013)
31. In-vivo multispectral fluorescent imaging. Technical report, Bruker BioSpin, November 2016
32. Product Note, IVIS Spectrum, advanced preclinical optical imaging. Technical report, PerkinElmer, Inc., 2016
33. Product Note, Solaris, advanced preclinical optical imaging. Technical report, PerkinElmer, Inc., 2016
34. Product Note, *Pearl Trilogy imaging system* (Technical report, Li-Cor, 2017)
35. B. Zhu, E.M. Sevick-Muraca, A review of performance of near-infrared fluorescence imaging devices used in clinical studies. Br. J. Radiol. **88**, 1–26 (2014)
36. J.M. Bioucas-Dias et al., Hyperspectral unmixing overview: geometrical, statistical, and sparse regression-based approaches. IEEE J-STARS **5**(2), 354–378 (2012)
37. W.K. Ma et al., A signal processing perspective on hyperspectral unmixing. IEEE Signal Process. **67**, 67–81 (2014)
38. N. Keshava, A survey of spectral unmixing algorithms. Linc. Lab. J. **14**(1), 55–78 (2003)
39. F. Cutrale, Hyperspectral phasor analysis enables multiplexed 5D in vivo imaging. Nat. Methods **14**(2), 149–155 (2017)
40. R. Heylen, P. Gader, M. Parente, A review of nonlinear hyperspectral unmixing methods. IEEE JSTAR **7**(6), 55–78 (2014)
41. L. Gao, R.T. Smith, Optical hyperspectral imaging in microscopy and spectroscopy - a review of data acquisition. J. Biophotonics **8**(6), 441–456 (2015)
42. S. Matteoli, M. Diani, J. Theiler, An overview of background modeling for detection of targets and anomalies in hyperspectral remotely sensed imagery. IEEE JSTAR **7**(6), 2317–2336 (2014)
43. J.C. Harsanyi, C.I. Chang, Hyperspectral image classification and dimensionality reduction: an orthogonal subspace projection approach. IEEE Trans. Geosci. Remote Sens. **32**(4), 779–785 (1994)
44. SPECTIR, SPECTIR: glossary of terms (2017), http://www.spectir.com/tools-resources/glossary-of-terms/. Accessed 29 June 2017
45. L. Frey, L. Masarotto, M. Armand, M.L. Charles, O. Lartigue, Multispectral interference filter arrays with compensation of angular dependence or extended spectral range. Opt. Express **23**(9), 11799–11812 (2015)
46. R. Levenson, J. Beechem, G. McNamara, Spectral imaging in preclinical research and clinical pathology. Anal. Cell. Pathol. **35**(5–6), 339–361 (2012)
47. X. Song et al., Automated region detection based on the contrast-to-noise ratio in near-infrared tomography. Appl. Opt. **43**(5), 1053–1062 (2004)
48. T. Sawyer, S.E. Bohndiek, Towards a simulation framework to maximize the resolution of biomedical hyperspectral imaging. Proc. SPIE Int. Soc. Opt. Eng. **10412**(104120C) (2017)
49. Occupation Health Services, University of Cambridge. Safe Use of Lasers University of Cambridge, HSD013R (rev 5) (2016)
50. T. Sawyer, A.S. Luthman, S.E. Bohndiek, Evaluation of illumination system uniformity for wide-field biomedical hyperspectral imaging. J. Opt. **19**(4), 045301 (2017)
51. A. Lambrechts et al., A CMOS-compatible, integrated approach to hyper- and multispectral imaging, in *2014 IEEE International Electron Devices Meeting, IEDM14* (2014), pp. 261–264
52. P.A. Valdes et al., Quantitative, spectrally-resolved intraoperative fluorescence imaging. Sci. Rep. **2**(798) (2012)
53. M. Jermyn et al., Macroscopic-imaging technique for subsurface quantification of near-infrared markers during surgery. J. Biomed. Opt. **20**(3), 036014 (2015)

54. S. Tzoumas, N.C. Deliolanis, S. Morscher, V. Ntziachristos, Unmixing molecular agents from absorbing tissue in multispectral optoacoustic tomography. IEEE Trans. Med. Imaging **33**(1), 48–60 (2014)
55. B.P. Joshi et al., Multimodal endoscope can quantify wide-field fluorescence detection of Barrett's neoplasia. Endoscopy **48**(2), A1–A13 (2015)

Chapter 4
A Multispectral Endoscope Based on SRDAs

Having established the potential of SRDAs to perform multiplexed fluorescence imaging in a wide-field system, we progressed to establish their utility for endoscopic imaging, relevant to cancer. The experience of using a SRDA in the wide-field set-up provided useful background data on how to address experimental complexities of imaging through a fibre bundle for endoscopy. Several studies have shown that simultaneous acquisition of spatial and spectral data can aid detection and diagnosis in biomedical and clinical imaging applications [1, 2]. These studies have generated a strong interest in combining spectral imaging of white light reflectance, providing contrast based on the absorption spectra of endogenous chromophores or by using exogenous fluorescent contrast agents [1, 2]. Here we combined spectral reflectance and fluorescence imaging in a bimodal multispectral endoscope designed with future clinical applications in mind. Two snapshot SRDAs were integrated in the bimodal endoscope to allow for reflectance imaging in the visible spectral region and multiplexed fluorescence imaging in the NIR.

This chapter presents the design, technical characterisation and initial imaging performance data of the bimodal multispectral endoscope. Multispectral endoscopic imaging and spectral unmixing of chemically oxy/deoxygenated mouse blood and three fluorescent dyes (AF647, AF660 and AF700) were demonstrated in a tissue mimicking agarose phantom. Detection of two fluorescent dyes, AF660 and 700, were additionally demonstrated in an ex vivo oesophageal porcine model. With further developments, this technology has the potential to improve early endoscopic cancer diagnosis, guide tissue biopsies and inform treatment choices [3]. Before detailing the bimodal multispectral endoscope developed here, this chapter opens with a discussion of the clinical need to improve endoscopic screening for GI cancers, a discussion of previously reported multispectral endoscopes, and a rationale for the use of SRDAs in endoscopy.

A. S. Luthman, *Spectrally Resolved Detector Arrays for Multiplexed Biomedical Fluorescence Imaging*, Springer Theses,
https://doi.org/10.1007/978-3-319-98255-7_4

4.1 Literature Review

4.1.1 Challenges of Early Cancer Detection in Endoscopy

Endoscopy is a clinical procedure that can be used to visualise the epithelial surfaces of hollow organs, commonly used to monitor and diagnose diseases of the GI tract. Cancerous lesions in the GI tract are conventionally detected and diagnosed via white light endoscopy (WLE). WLE detects cancerous lesions based on structural changes or discolouration of the epithelial surface [3], and may also be used to guide the acquisition of tissue biopsies [4]. Early detection of GI cancers is critical to improve the prognosis [3]. However pre-cancerous lesions are often highly focal and patchy in nature, with a flat architecture, making them difficult to detect with conventional WLE [3, 4]. It has been reported that up to 26% of grossly visible adenomas, and 27% of all dysplastic GI lesions have a flat architecture [3]. Additionally, even after the successful identification of a lesion and a subsequent biopsy, the subjectivity of the pathologist often introduces an uncertainty regarding the confirmation and staging of the dysplastic lesion [4]. Hence, there is a great need to improve diagnostic performance of endoscopic cancer screening of the GI tract.

This is of particular relevance to patients with Barrett's Oesophagus. Barrett's Oesophagus is a change in the oesophageal epithelium arising from chronic inflammation of the oesophageal surface [4]. Patients with Barrett's have a 12–40% enhanced risk of developing oesophageal adenocarcinoma (EAC) via stages of low and high grade dysplasia (LGD and HGD) [5]. Whereas the absolute risk of developing EAC from Barrett's is low (0.5%), the incidence of EAC in patients with Barrett's is significantly higher than in the general population [4]. It is therefore recommended that these patients undergo routine endoscopic surveillance every three to four years to allow early cancer diagnosis [6], which is key to improved treatment outcome. Early diagnosis of EAC is critical since resection [7], or radiofrequency ablations [8], of early stage dysplastic lesions may be curative [9]. Early detection rates via conventional WLE surveillance, paired with biopsies according to the Seattle protocol (which prescribes the acquisition of four random quadrant biopsies every 2 cm), are however low [4], due to high biopsy miss rates, low adherence to the biopsy protocol, and subjectivity of diagnosis [4]. Consequently, there is strong interest in improved optical imaging methods that provide endoscopic screening with enhanced diagnostic sensitivity and that may reduce the number of required biopsies [3–5, 10–14].

4.1.2 Advanced Endoscopy Based on Intrinsic Tissue Contrast

Several methods with improved endoscopic tissue contrast, beyond that of WLE are already implemented into clinical use. Many of these methods, such as narrowband imaging (NBI), AF imaging (AFI) endoscopy and trimodal imaging, rely on the acquisition of spectrally resolved image data. AFI allows visualisation of intrinsic tissue fluorescence within the 500–630 nm spectral region [4]. Here tissue fluorescence, excited with ultraviolet and blue light (390–470 nm), is combined with tissue visualisation via green light reflectance imaging [4]. This yields a real-time pseudocolour image where healthy tissue appears green and dysplastic tissue purple [4]. NBI increases visualisation of vascularisation and mucosal surface patterns and thereby may reveal highly vascularised tumours by reflectance imaging with blue (440–460 nm) and green light (540–560 nm), matched to the absorption spectra of haemoglobin [4, 10]. In trimodal imaging, AFI and NBI are combined with high definition WLE for increased tissue contrast [4]. Although not strictly relaying on intrinsic tissue contrast, chromoendoscopy is also frequently used to enhance the contrast between cancerous lesions and the healthy epithelium via applications of non specific stains such as acetic acid, indigo carmine and methylene blue [4].

All of the clinically implemented endoscopy techniques are, however, somewhat limited. Whereas both AFI and NBI can increase the detection of dysplastic lesions and effectively guide the acquisition of biopsies [6], these methods have difficulty in distinguishing between inflamed tissue and neoplastic lesions [3]. Identification of neoplastic lesions is harder on inflamed tissue since inflammation may alter the vascular pattern and AF properties of tissue [15]. Both AFI and NBI also suffer from high false positive rates [3, 4, 13]. NBI is however recommended by the American Society for GI Endoscopy (ASGE) for surveillance of patients with Barrett's Oesophagus. Endoscopic NBI has in meta-analysis studies been shown to exceed the required sensitivity (>90%), negative predictive value (>98%) and specificity (>80%) thresholds for detection of HGD and early EAC set by the ASGE. When NBI and AFI are combined together with high definition WLE—as in trimodal imaging—the detection rate of both HGD and AEC is further increased [10]; unfortunately these results do not extend to the detection of early dysplasia [10].

Chromoendoscopy with methylene blue or indigo carmine shows limited performance improvement over WLE [6]. In contrast, chromoendoscopy with acetic acid is recommended by the ASGE to guide the acquisition of biopsies during endoscopic surveillance of patients with Barrett's Oesophagus [6]. The acetic acid reacts with the oesophageal epithelial surface causing it to whiten; thereby the dysplastic lesions can be identified as they regain their colour quicker than the surrounding epithelium [16]. Although acetic acid guided biopsy needs to be further verified before replacing the current standard biopsy protocol [16], these results highlight the potential benefits of using external contrast agents during endoscopic screening.

To further increase the intrinsic tissue contrast, several research groups are extending the spectral acquisition range beyond the clinically used techniques. Increasing the number of spectral bands, in combination with data analysis using unsupervised spectral unmixing algorithms, have shown promise to improve the visualisation of biomarkers such as the vascular pattern and the oxygenation status of blood [17–19], as well as for improved detection of gastric [20] and colorectal cancerous lesions [21–23]. Many of these systems do however rely on acquisition of spectral data via tunable filters [21] or filter wheels [18–20, 22], such that temporal resolution is sacrificed for the acquisition of spectral data. These systems are also in relative early stages of development, and thus will require further clinical evaluation.

4.1.3 Targeted Contrast Agents and Optical Molecular Imaging

There is a strong clinical interest in optical molecular imaging for endoscopic diagnostic cancer screening of the GI tract [3, 5, 14]. Optical molecular imaging relies on molecular probes that attach to specific protein targets (either on the cell surface, in the cytoplasm or in the extracellular matrix) which may be over or under expressed in cancers [3, 5]. Typically, an optical molecular probe consists of a reporter, such as a fluorescent dye, conjugated to a targeting moiety which binds specifically to the target protein. This makes it possible to visualise the protein via fluorescence emitted by the reporter. For endoscopic cancer screening, such molecular probes may be topically applied with a spray catheter threaded through the accessory channel of a clinical endoscope [10]. Topical application of molecular probes is particularly well suited to GI cancers since dysplasia frequently originates in the epithelial layer [3]. Topical application of molecular probes also allows analysis by contrast enhanced imaging with minimal disruption to the clinical work-flow and reduced exposure to potential probe toxicity with minimal systematic bio-distribution [14].

Optical molecular imaging has many potential benefits over imaging methods which rely on non-specific contrast. For example, specific molecular fluorescence read-out may in the future enable objective diagnostic read-out via fluorescence threshold images (rather than the current subjective interpretation of histology slides) [4], diagnosis based on molecular changes occurring before structural changes [5], and personalised therapy and chemoprevention [3]. Several types of molecular probes, with different fluorescent dyes and targeting moieties, are currently under preclinical [3, 24] and clinical investigation [3, 10, 25]. In clinical studies, the application of a fluorescence molecular contrast agents have been evaluated as a 'red-flag', prompting tissue biopsy for colorectal [26] and oesophageal cancer [10]. These early studies have indeed indicated that optical molecular imaging in endoscopy may reduce the number of required biopsies and improve the early detection rates of dysplastic lesions [10, 26].

The non-uniformity of contrast agent spraying and the heterogeneity in biomarker expression are however frequently cited as significant challenges for optical molecular imaging [3, 5, 14, 27]. These challenges may be overcome by the acquisition of spectral image data. For example, multiplexed fluorescence imaging of several molecular probes may be required for effective cancer screening of diverse patient populations with varying levels of biomarker expression [5]. Simultaneous spraying of a non-targeted control and a targeted molecular fluorescent contrast agent may allow for real-time correction of spraying non-uniformities. A paired probe approach could be particularly useful for contrast agents with a negative binding response, such as the lectin based molecular probe for endoscopic screening of patients with Barrett's Oesophagus developed by Bird-Lieberman et al. [24]. Previously reported clinical studies of endoscopic molecular imaging have utilised bimodal endoscopes, with separate WLE and fluorescence channels to record white light images and to detect one fluorescent dye [10, 26]. There is however a strong interest in instrumentation advances enabling clinical multiplexed fluorescence endoscopy [3, 5].

4.1.4 Existing Approaches to Spectral Endoscopy

Spectral endoscopy is a newly emerging field, and relatively few systems have been reported in the literature. As previously noted, many of the systems developed for molecular imaging are not fully MSI/HSI systems [10, 26]. Instead, a dedicated fluorescence channel has been added to a standard WLE to detect the fluorescence signal from a single molecular probe [10, 26].

Although MSI/HSI endoscopes for molecular imaging have yet to make it to clinical studies, several systems have been described in the literature. These systems are reviewed here; since we are aiming towards GI endoscopy this review is focused on flexible spectral endoscopes. A majority of the reported flexible MSI endoscopes operates on the classical babyscope model, i.e. the threading of a flexible coherent fibre bundle (CFB) for MSI data collection through the accessory channel of a larger clinical endoscope [14]. Broadband light is typically coupled to the illumination port of the CFB, and spectral separation of the light performed at the back-end of the endoscope [14, 26, 28–30]. Spectral filtering is generally performed via amplitude/image or optical path division methods. Spectral imaging has, for example, been achieved via a set of dichroic beamsplitters and dedicated detectors [14, 26, 29], a filter wheel with a set of bandpass filters [30] and a tunable filter [28]. Spectral imaging can also be achieved by filtering the input light. Examples include filtering broadband input light by a wheel of bandpass filters [22, 31], a tunable filter [21] and a monochromator [32]. Alternatively, a set of laser lines may be used to perform spectral imaging [11, 29, 33]. Due to the sequential acquisition of spectral image data, spectral data are often acquired at the expense of temporal resolution. Spectral image acquisition via path division methods does not sacrifice temporal resolution. These system do instead include multiple dedicated detectors and dichroic filters [11, 33], making the systems both bulky and expensive.

The scanning fibre endoscope (SFE) system is based on rapid sequential spatial scanning of the FOV, whilst acquiring data in 4 spectral bands. Three R-G-B laser lines are coupled into a waveguide which is vibrated at mechanical resonance to spatially scan the full FOV in an outwardly growing spiral [11]. The RGB lasers can be used to collect confocal colour images over a large FOV [11]. The endoscope can also be adapted to fluorescence imaging by collecting laser excited intrinsic or extrinsic fluorescence signals with a concentric ring of high NA fibres surrounding the vibrating waveguide [11]. Spectral separation of the light is performed by a set of dichroic filters at the distal end of the SFE [11]. Although spectral imaging is achieved in a conventional manner, the active scanning allows for imaging over a large FOV with high spatial resolution, reconfigurable image properties and concurrent spectral image acquisition [11]. SFE has indeed been used for concurrent endoscopic imaging of three fluorescently labelled peptides in a mouse model, targeted towards colorectal adenomas [34].

Extraction of spectral data via spatial scanning can also be achieved by spectrally encoding the spatial information [35]. For example, Zeidan and Yelin [35] integrated a prism and diffraction grating on the distal tip of a rigid endoscope, such that broadband light (400–700 nm) was spectrally dispersed in a line profile along the endoscopic axial direction and each spatial position illuminated by a different wavelength in each image frame. As a result, the full HSI data cube of the FOV could be be reconstructed by gradually rotating the rigid endoscope during endoscopic withdrawal. Impressively, the prototype endoscope was able to spectrally detected a burn mark inside of a turkey artery. However, sophisticated software algorithms have to be developed for real-time visualisation of the spectral data, and for application in less well controlled clinical environments.

A couple of spectral endoscopy methods which do not require either temporal or spectral scanning have also been proposed. For example, image mapping spectroscopy (IMS) has been adapted for endoscopy by using the *babyscope* method, combined with compact IMS hardware, to spatially remap, spectrally disperse and image the distal end of the CFB [17]. The performance of this endoscope has however yet to be demonstrated in a clinical application. Another method which side-steps the need for spatial and spectral scanning is the "4-D endoscope" presented by Lim and Murukeshan [36]. Here, the CFB is reformatted into a linear fibre array and directly coupled to the input port of a spectrometer. This allows the spectral range between 400–1000 nm from each fibrelet to be densely spectrally sampled. The endoscope's spatial resolution is however currently limited to 100 spatial points. The spatial resolution is presumably limited by instrumentation challenges associated with reformatting the CFB. Moreover, the utility of the "4-D endoscope" has not yet been evaluated in a clinical setting.

4.1.5 The Potential of SRDAs for Endoscopy

Clinical implementations of spectral endoscopes are currently held back by complex data analysis and interpretation, as well as the high cost and complexity of the equipment [5]. Several of the previously reported spectral endoscopy systems include multiple bandpass filters, several laser lines, or multiple detectors dedicated to separate spectral bands; the use of multiple expensive optical components make the systems both bulky and costly. Here we have instead sought to realise spectral endoscopy via snapshot SRDAs. As discussed in Sect. 2.1.1, SRDAs are highly suitable for clinical imaging applications due to their compact and robust nature and potential for very cost effective manufacturing.

Snapshot SRDAs allow for simultaneous acquisition of image data in several spectral bands, such that no image co-registration is required to correct motion artifacts arising due to time-offset between acquisitions in different spectral bands. Concurrent spectral data acquisition provides a significant advantage for image co-registration [20]. Image co-registration typically relies on the identification of salient features, which is a non-trivial task for moist and highly homogeneous tissues such as the GI tract [20]. SRDAs also have regular spectral filter band mosaic patterns which allow straightforward extraction of spectral data. The potential to rapidly acquire and analyse spectral data may allow for real-time superposition of spectral information on traditional WLE images. Direct visualisation of fluorescence data on WLE images also has potential to simplify data analysis.

Here we integrated two snapshot SRDAs in a bimodal endoscope to allow for reflectance imaging in the visible spectral region, and multiplexed fluorescence imaging in the NIR. This approach allow the simultaneous extraction of intrinsic and extrinsic tissue contrast. The endoscope system is assembled around a commercial modular endoscope (PD-PS-0095, PolyScope®; PolyDiagnost). This has many advantages for translation to the clinical setting, for example, the small diameter of the modular endoscope (2.65 mm outer diameter; 3 mm atraumatic cap at the distal end). The small diameter of the modular endoscope allows it to be operated through the working channel of a larger endoscope [37]. This can allow future direct performance comparison to the current standard-of-care [37]. Its availability under CE marking is also an advantage [37].

From our previous work with SRDAs in a wide-field system, we learned that the reflectance-detection geometry set-up typically encountered in endoscopy limits the DR of the camera when performing fluorescence imaging. For the integration of SRDAs in endoscopy, we therefore chose to include a dichroic and notch filter in the optical beam path to separate the reflectance and fluorescence light. This allowed us to increase the intensity of the reflectance and fluorescence excitation illumination without saturating the SRDA in the fluorescence imaging channel. The endoscope's potential for concurrent reflectance and fluorescence imaging was demonstrated by simultaneously imaging and unmixing chemically oxy/deoxygenated mouse blood and three fluorescent dyes in a tissue mimicking phantom. Detection of two fluorescent dyes was also demonstrated in an ex vivo oesophageal porcine model.

4.2 Experimental Methods

4.2.1 Endoscope Design and Assembly

A bimodal endoscope (Fig. 4.1), incorporating the Imec visible and NIR snapshot SRDAs, was assembled to allow simultaneous visible multispectral reflectance and NIR multiplexed fluorescence imaging in an endoscopic set-up. Light from a white light broadband LED and a NIR laser diode was used to simultaneously illuminate

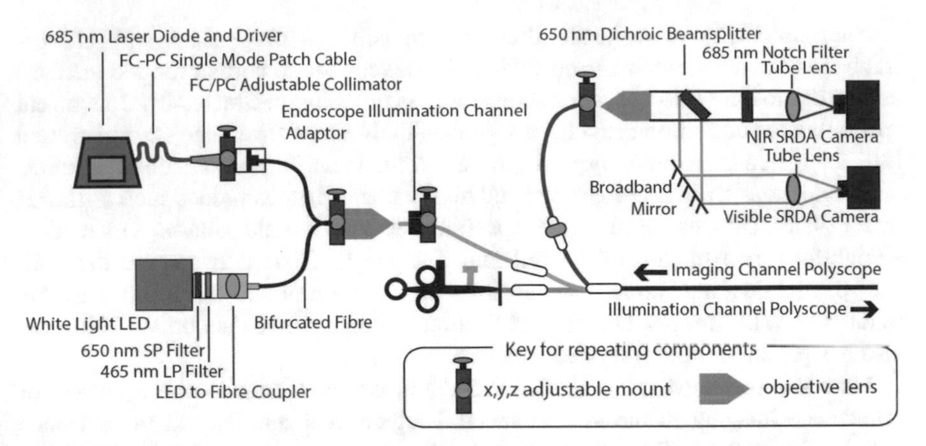

Fig. 4.1 Multispectral bimodal endoscope design, incorporating the Imec visible and NIR snapshot SRDAs, assembled to allow simultaneous visible multispectral reflectance and NIR multiplexed fluorescence imaging in an endoscopic set-up. Light from a white light broadband LED and a NIR laser diode simultaneously illuminate the sample during imaging to provide light for reflectance imaging in the visible, and excitation of fluorescent dyes in the NIR. A dichroic beamsplitter on the back-end of endoscope directs reflectance and fluorescence light towards the visible and NIR snapshot sensor respectively. The endoscope is assembled around a commercial modular endoscope (PolyScope®) to allow future translation to clinical endoscopic imaging applications

the sample during imaging to provide light for reflectance imaging in the visible and excitation of fluorescent dyes in the NIR. The endoscope was assembled around a commercial modular endoscope (PolyScope®) to allow future translation to clinical endoscopic imaging applications. The optical components were selected based on a model of the spectral transmission characteristics of the endoscope detailed in Sect. 4.2.2.

Light from a white light broadband LED and a NIR laser diode were coupled into the illumination port of the Polyscope® via a bifurcated fibre. Light from the white light LED (T7358; Prizmatix) was filtered with a short and longpass filter (AT465lp; Chroma and FESH0650; Thorlabs) before coupling into one leg of the bifurcated fibre (19 Fibre Y bundle, BF19Y2HS02; Thorlabs) using a custom LED-to-SMA coupler (Prizmatix). The laser diode (LP685-SF15; Thorlabs) was placed in an integrated mount with current and temperature control (CLD1011LP; Thorlabs), and coupled into the second leg of the bifurcated fibre via a FC-PC single mode patch cable (P1-630A-FC-1; Thorlabs) and an adjustable collimator (CFC-5X-B; Thorlabs) mounted on an x–y–z adjustable mount (CXYZ1/M; Thorlabs). The collimator was selected such that the divergence angle and mode field diameter (MFD) of the focused beam matched the NA and diameter of the bifurcated fibre. Coupling of laser light into the bifurcated fibre was performed by placing a photodetector (1916-R; Newport) at the output port of the fibre whilst adjusting the focus and positioning of the collimator to maximise the transmitted light. An objective lens (40X/0.65 Plan Achromat Objective, RMS40X; Olympus) was used to couple light from the output port of the bifurcated fibre into the illumination channel of the PolyScope® (PD-PS-0095, PolyScope®; PolyDiagnost). The objective lens was selected based on its high magnification and good match to the PolyScope® illumination channel NA of 0.63.

The bifurcated fibre, objective lens and illumination channel of the PolyScope® were cage mounted to allow for efficient coupling of light into the illumination channel of the PolyScope®. To cage mount the PolyScope® illumination channel, a custom 3D printed mount was made to slot the illumination channel into an adapter and cage mountable plate (AD16F and CXY1; Thorlabs). The coupling efficiency between the bifurcated fibre and the PolyScope® illumination channel was optimised by adjusting the position of the bifurcated fibre, objective lens and illumination channel whilst monitoring the power at the distal end of the endoscope. The final power on sample was 1.74 ± 0.01 mW for the laser line and 1.60 ± 0.01 mW for the broadband LED.

On the back-end of the endoscope, the CFB of the PolyScope® imaging channel was imaged by the visible and NIR SRDAs. An infinity corrected objective lens (Plan Fluorite 20X, UPLFLN20X; Olympus) was used to magnify the image of the CFB. A dichroic beamsplitter (652 nm single-edge dichroic, FF652-Di01; Semrock) was then used to direct reflectance and fluorescence light towards the visible and NIR SRDA respectively. A notch filter (685 nm Laser Notch Filter, ZET685NF; Chroma) was placed in the fluorescence beam path to prevent laser light from saturating the NIR SRDA. Two 100 mm focal length tube lenses (Air-Spaced Achromatic Doublet, AR Coating 350–700, ACA254-100-A, and 650–1050 nm, ACA254-100-B; Thorlabs) were used to image the CFB onto the two SRDAs. A corner mounted broadband mirror (21015; Chroma) was included in the reflectance channel for ease of alignment and to increase the system compactness. The complete back-end of the endoscope was mounted in a cage system with x–y–z control for easy alignment.

The fluorescence and reflectance channels of the endoscope were separately aligned. An external light source (OSL2 with OSL2BIR bulb; Thorlabs) was used during alignment, since the system was designed to block reflectance light from reaching the NIR SRDA. Firstly, the dichroic beamsplitter was removed from the imaging channel to direct all light towards the NIR SRDA. The position of the objective lens was adjusted until a bright circle could be imaged with the camera. A target with sharp features was then placed under the distal tip of the endoscope. The relative position of the objective lens and the CFB of the PolyScope's® imaging channel was adjusted until a focused image of the target could be observed. To complete the alignment of the fluorescence channel, the relative distance between the tube lens and the SRDA was fine adjusted to enhance the image focus. After successful alignment of the fluorescence channel, the dichroic beamsplitter was slotted back into place. The process was repeated for the reflectance channel of the endoscope, also adjusting the tilt of the cage corner mounted mirror. The alignment was completed by ensuring that the externally illuminated target could be simultaneously imaged in the reflectance and fluorescence channel of the endoscope.

Because of the low QE of the SRDAs (Fig. 2.2, Sect. 2.1.3), optics were selected to produce low magnification images of the CFB to maximise signal intensity. Relatively short focal length lenses were used in order to keep the system compact whilst also

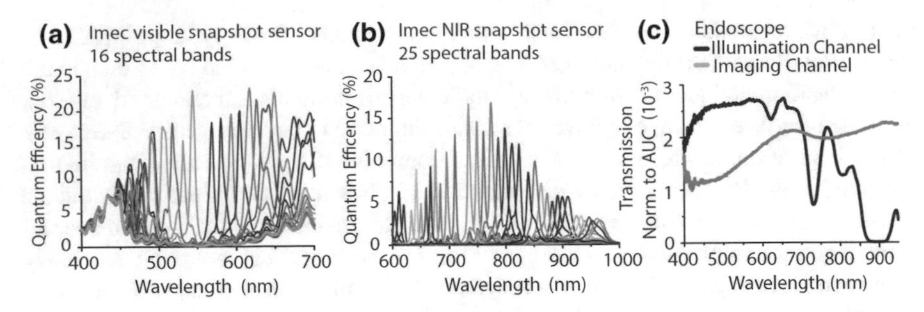

Fig. 4.2 The spectral transmission characteristics of the employed optical components should be considered when designing a spectral endoscope. Optical components were therefore selected and matched to the spectral characteristics of key components of the endoscope to optimise the detected signal intensity and to minimise cross-talk between the reflectance and fluorescence imaging channels of the endoscope. Optical components were matched to the spectral response of **a** the Imec visible snapshot SRDA, **b** the Imec NIR snapshot SRDA, and **c** the spectral transmission of the illumination and imaging channels of the Polyscope®. The QE curves of the SRDAs were provided by the supplier whereas the transmission spectra of the Polyscope® illumination and imaging channels were measured in-house

imaging at a high F/#. In contrast to the wide-field fHSI system (Chap. 3), the light collection efficiency of the endoscope was mainly determined by the NA of the imaging channel of the PolyScope® and it was therefore preferable to image at a high F/# to ensure efficient light coupling and narrow spectral sampling by the SRDAs' spectral filters.

The system was mounted in a black box (TB5 Black Posterboard, XE25L225/M, XE25L375/M Construction Rails and XE25W3 Quick Corner Cube; Thorlabs) on a optical breadboard (MB4545/M; Thorlabs) for ease of transport and use.

4.2.2 Model of the Endoscope's Spectral Transmission Characteristics

The spectral transmission characteristics of the employed optical components should be considered when designing a spectral endoscope. The optical components and the fluorescent dyes used in this study were therefore selected to best match the spectral characteristics of the three key components of the endoscope; the Imec visible snapshot SRDA (Fig. 4.2a), the Imec NIR SRDA (Fig. 4.2b), and the spectral transmission of the PolyScope® (Fig. 4.2c). A MATLAB® model was developed to predict the spectral characteristics of the full endoscopic illumination and imaging channel and the expected fluorescence signal detected, when using a particular set of optical components. The results of the model were used to guide the selection of optical components and fluorescent dyes used in this study. The optical components and fluorescent dyes were selected to; optimise the overlap between the illumination and the fluorescence absorption, maximise the detected fluorescence signal, and minimise cross talk between the reflectance and fluorescence imaging channels. Optical components tested in the model were initially selected to produce a relatively low-cost, compact and robust endoscopic system.

The model also served a dual purpose; it was also used to calculate spectral endmembers for unmixing of the oxy/deoxygenated blood and the fluorescent dyes imaged in this study. In many clinical applications, direct measurements of the spectral endmembers are not possible and it can therefore be advantageous to determine the spectral endmembers via modelling. The model may be used to calculate spectral endmembers by propagating reference spectra through the modelled spectral response of the endoscope's reflectance and fluorescence imaging channels.

After assembly of the spectral endoscope, the accuracy of the model and the endmember predictions were evaluated via experimental measurements. Whereas the model, the predictions from the model and the acquisition of experimental data are described in the following section, the comparison between the modelled and experimentally measured spectra are presented in Sect. 4.3.1.

Model of the Illumination and Imaging Channels

The optical components of the bimodal endoscope were selected by considering the resultant spectral transmission of the illumination and imaging channels of the endoscope. To select optical components, the illumination and imaging channels of the endoscope were modelled in MATLAB® 2015 prior to system assembly. We followed the established method of modelling the spectral transmission characteristics of an optical system's response function by linear multiplication of the transmission functions of its individual components [38].

The majority of the spectral transmission data of the optical components were obtained in tabular form from suppliers, or extracted from supplier provided reference plots with the user defined MATLAB® function GRABIT [39]. Transmission spectra not available from the suppliers were experimentally measured. The transmission spectra of the illumination and imaging channel of the PolyScope® were measured by butt coupling a stabilised broadband light source (SLS201; Thorlabs) to the illumination and imaging channel of the PolyScope® and the modulated broadband output spectra measured at the distal and proximal end of the PolyScope® using a spectrometer (AvaSpec-ULS2048-USB2-FCDC; Avantes). The acquired spectra were divided by the known spectrum of the broadband light source to obtain the transmission properties of the illumination and imaging channels of the PolyScope®. Spectra of broadband LEDs, considered to be used as endoscopic reflectance light sources, were measured by coupling the inspection fibre from the spectrometer to the LED-to-SMA coupler, mounted at the output ports of the LEDs. Neutral density (ND) filters (NDK01 ND filter set; Thorlabs) were added to the coupler to prevent saturation of the spectrometer; the spectral effect of the ND filters were removed by division.

The overall transmission characteristics of the illumination (Fig. 4.3a) and imaging channels of the endoscope (Fig. 4.3b) were predicted by multiplying the normalised transmission spectra of the individual components in the endoscope's illumination

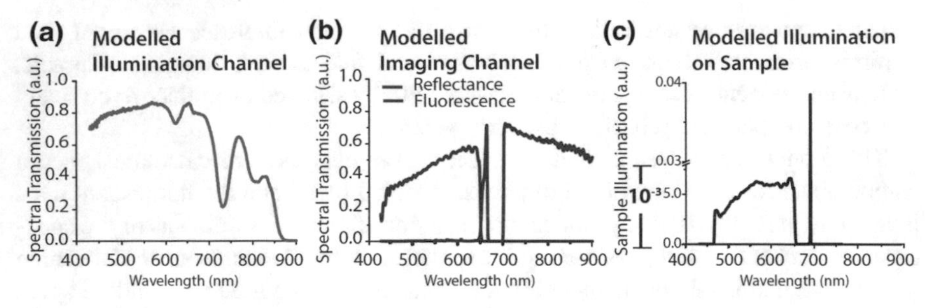

Fig. 4.3 The modelled overall transmission spectra of the **a** illumination and **b** the reflectance and fluorescence imaging channels of the endoscope. Distinct spectral features of the imaging channels include the effect of a 650 nm dichroic beamsplitter directing reflectance and fluorescence light respectively towards the visible and the NIR SRDA, and a drop in the transmission of the fluorescence imaging channel caused by a 685 nm notch filter in the fluorescence imaging channel. **c** The illumination at the distal tip of the endoscope was modelled by linearly adding and propagating the power spectra of the selected broadband LED and diode laser through the modelled transmission spectra of the endoscopic illumination channel. The transmission and illumination spectra shown were modelled based on the optical components included in the final endoscope design; the illumination spectrum was normalised to AUC

and imaging channel. A calculation was additionally performed to approximate the illumination spectrum at the distal tip of the endoscope (Fig. 4.3c).

The bimodal endoscope was designed to have two separate light sources; a white light broadband LED for reflectance imaging and a laser line for fluorescence excitation. The LED and laser line were coupled into separate legs of the bifurcated fibre coupled to the illumination port of the PolyScope®; this was designed to provide simultaneous reflectance and fluorescence sample illumination. The combined illumination spectrum $I_{ill}(\lambda)$ was approximated by linearly adding the power spectra of the LED and the laser, whilst also approximating the effects of coupling losses and the spectral transmission of components prior to the bifurcated fibre;

$$\begin{aligned} I_{ill}(\lambda) &= A_{fibre} \times I_{LED}(\lambda) \times T_{SP}(\lambda) \times T_{LP}(\lambda) \\ &+ \frac{P_{laser}}{AUC_{laser}} I_{laser}^{norm} \times 0.63 T_{patch\ cable}(\lambda) \times 0.4 T_{collimator}(\lambda), \end{aligned} \tag{4.1}$$

where I_{LED} is the experimentally measured intensity spectra of the white light LED after the SMA coupler, multiplied by the transmission spectra of the shortpass $T_{SP}(\lambda)$ and longpass filters $T_{LP}(\lambda)$ placed in the LED-to-SMA coupler. To approximate the total LED light coupled into the bifurcated fibre, the LED intensity spectrum was multiplied by the total fibre area in the leg of the bifurcated fibre coupled to the LED (A_{fibre}). The laser power spectrum was approximated by multiplying the normalised laser output spectra (I_{laser}^{norm}), provided by the suppliers, by the ratio of the supplier quoted laser power (P_{laser}) and the total AUC (AUC_{laser}) of the normalised laser spectra. The effect of coupling optics prior to the bifurcated fibre was taken into account by multiplying the calculated laser power spectra by the normalised transmission spectra of the patch-cable ($T_{patch\ cable}(\lambda)$) and the collimator ($T_{collimator}(\lambda)$). Coupling losses were approximated by multiplying the 'worst-case-scenario' transmission loss of the patch cable (0.63) with the ratio between the total fibre area in the bifurcated fibre and the expected beam MFD (0.4). The modelled LED and laser power spectra were linearly added and normalised to AUC; the calculation was not used to model the absolute power on sample as the coupling losses in the common optical path of the laser and LED were not taken into account in the calculation. The normalised spectrum was subsequently propagated through the common illumination channel of the endoscope to approximate the spectral distribution of the sample illumination.

Model of the Detected Fluorescence and Reflectance Signals

The fluorescence signal from five fluorescent dyes (AF647, AF660, AF680, AF700, AF750, AF780; Invitrogen) were modelled to select a subset of dyes whose absorption and emission spectra were well matched to the sample illumination, the spectral transmission of the fluorescence channel and the detection efficiency of the NIR SRDA.

The expected fluorescence signal of each dye ($I_{em}(\lambda)$) was modelled as;

$$I_{em}(\lambda) = I_{em}^{norm} \times \varepsilon^{+} \times QE_{dye} \times \int I_{sample}^{norm}(\lambda) \times I_{ex}^{norm} d\lambda, \tag{4.2}$$

where $I_{sample}^{norm}(\lambda)$ is the modelled sample illumination, I_{ex}^{norm} the excitation and I_{em}^{norm} the emission spectrum of the fluorescent dye (all normalised to AUC). The extinction coefficient (ε^{+}), the QE (QE_{dye}) and the emission and absorption spectra of the fluorescent dyes were obtained from the supplier. Here we assume that the fluorescence intensity was linearly proportional to the light available for fluorescence excitation, the extinction coefficient and the QE of the dye. It should be noted that, whereas linear proportionality may be assumed in low absorbance solutions or suspensions [40], this naive model does not consider the wavelength dependence of the extinction coefficient. The extinction coefficient is typically defined at the peak wavelength of the excitation spectra [40]; its value may therefore not be accurately represented in the current model. Nevertheless, this simplistic model was considered sufficient to guide the selection of dyes for multiplexed fluorescence imaging.

To compare the fluorescence signals detected in the reflectance and fluorescence imaging channels, the fluorescence light incident and sampled by the visible (Fig. 4.4a, b) and NIR (Fig. 4.4c, d) SRDAs were calculated. The emission spectra of the dyes were propagated through the predicted transmission spectrum of the endoscope's reflectance and fluorescence imaging channels, and the light sampled by SRDAs ($I_{band\ :\ detected}$) modelled as;

$$I_{band\ :\ detected} = \int I_{cam}(\lambda) \times T_{band}(\lambda) d\lambda, \tag{4.3}$$

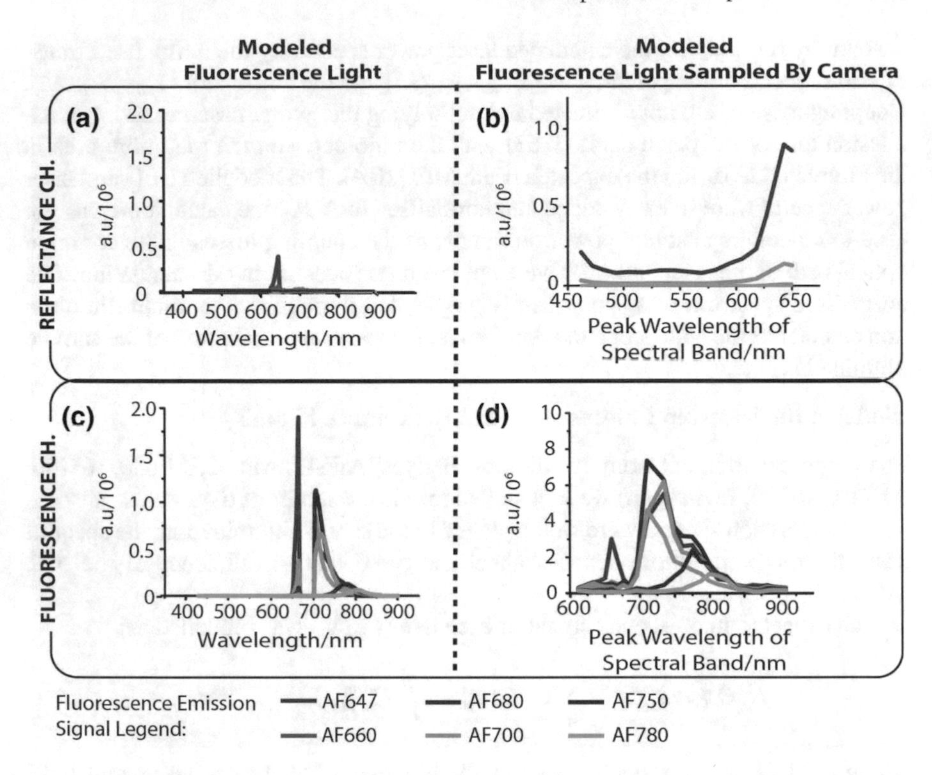

Fig. 4.4 The modelled fluorescence signal (**a**) incident on and **b** sampled by the 16 spectral bands of the visible SRDA, in comparison to that **c** incident on and **d** sampled by the 25 spectral bands of the NIR SRDA. The incident and detected fluorescence signals are expressed in arbitrary units, normalised such that the fluorescence signals may be compared across the two channels. The model aimed to guide the selection of optical components to maximise the fluorescence signals measured by the NIR SRDA, whilst minimising crosstalk between the reflectance and fluorescence imaging channels. The fluorescence signals have here been modelled based on the optical components selected for the final endoscope design

where $T_{band}(\lambda)$ is the supplier provided combined QE of a spectral filter band and the underlying image sensor, and $I_{cam}(\lambda)$ the modelled fluorescence signal incident on the SRDA. The distinct fluorescence signal detected in each of the 16 and 25 spectral bands of the visible and NIR SRDAs respectively, resulted in discretised modelled spectra, akin to that expected from experimental measurements with the SRDAs. The calculation was repeated for the modelled reflectance light to predict the reflectance light incident on (Fig. 4.5a) and sampled (Fig. 4.5b) by the visible and NIR SRDAs (Fig. 4.5c, d). A uniformly reflecting target was assumed when modelling the reflectance light.

The model allowed prediction of the spectral effect of an optical component prior to system assembly. This provided a flexible model to guide the selection of the optical components and fluorescent dyes used in this study; a set of optical component combinations were tested and the results of the model studied to select

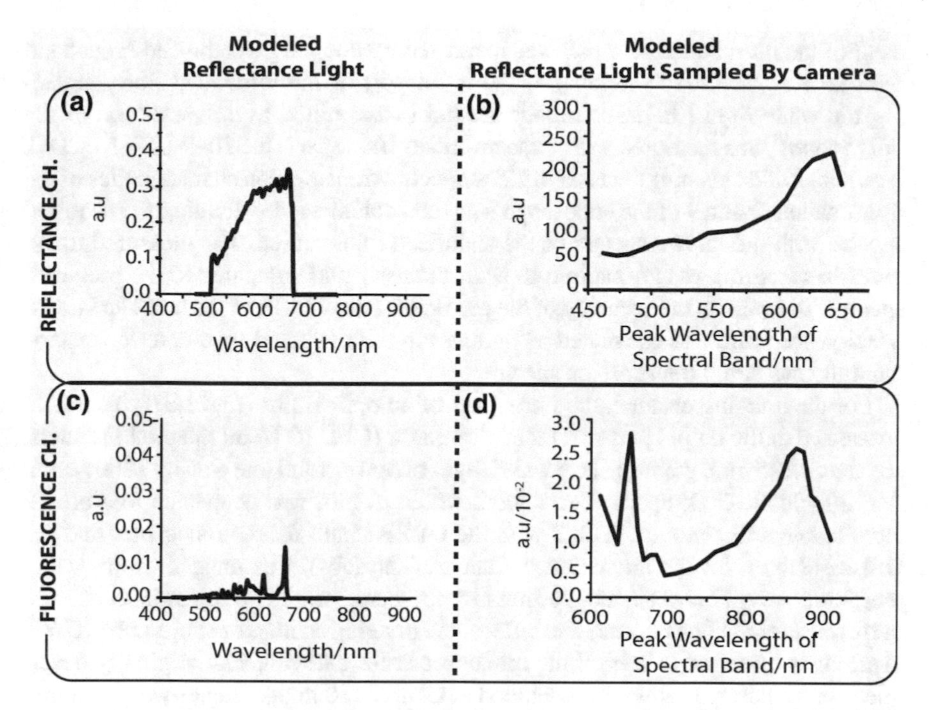

Fig. 4.5 The modelled reflectance signal (**a**) incident on and **b** sampled by the 16 spectral bands of the visible SRDA, in comparison to that **c** incident on and **d** sampled by the 25 spectral bands of the NIR SRDA. The incident and detected reflectance signals are expressed in arbitrary units, normalised such that the signals may be compared across the two channels. The model aims to guide the selection of optical components to maximise the reflectance signal measured by the visible SRDA, whilst minimising crosstalk between the reflectance and fluorescence imaging channels. The reflectance signal has here been modelled based on the optical components selected for the final endoscope design

the components which maximised the detected fluorescence signal and minimised cross talk between the reflectance and fluorescence imaging channels. The optical components tested in the model are listed in Appendix A; the ultimately selected components have been previously detailed in Sect. 4.2.1.

In addition to optimising the endoscope design, the model was also used to calculate spectral endmembers for unmixing of the oxy/deoxygenated blood and the fluorescent dyes imaged in this study. Endmembers for spectral unmixing can be calculated by propagating reference spectra through the modelled spectral response of the endoscopic imaging channels; this may be advantageous in cases where direct experimental measurements of the spectral endmembers are not possible.

Experimental Evaluation of Model Accuracy: Spectral Transmission Characteristics

The accuracy of the model of the spectral transmission characteristics of the endoscope was determined by experiment after system assembly. The transmission spectrum of the illumination channel was measured by coupling a stabilised broadband light source (SLS201; Thorlabs) to the input port of the bifurcated fibre, replacing the white light LED. The broadband light transmitted to the distal end of the PolyScope® was measured with a spectrometer (AvaSpec-ULS2048-USB2-FCDC; Avantes), whilst keeping the laser off. The spectral transmission characteristics of the illumination channel of the endoscope was then calculated by dividing the acquired spectra with the known spectra of the stabilised light source. The measured transmission spectrum was normalised to peak intensity and compared to the predicted spectra. To evaluate the accuracy of the prediction, the RMSE of the modelled transmission spectrum was calculated by comparing the modelled spectra to the experimentally measured transmission spectra.

For the imaging channel, the fibre facet of an optical fibre (M71L01; Thorlabs) connected to the output port of a monochromator (CM110 1/8m; Spectral Products, slit size: 0.125 mm, grating: blaze wavelength of 500 nm and line density 1200 g/mm, AG1200-00500-303; Spectral Products), which in turn was connected to a broadband halogen light source (OSL2 with the OLSB2 bulb for the reflectance and the OSL2BIR bulb for the fluorescence channel; Thorlabs), was imaged as the wavelength was swept between 400–900 nm in 3 nm increments. Two separate scans were performed for the fluorescence and reflectance imaging channels using a LabVIEW® visual user interface to control the monochromator and cameras, originally developed for the camera calibrations detailed in Chap. 2. 10 image frames were acquired at each wavelength increment. The integration times of the visible and NIR camera were set to 16 and 60 ms respectively. The gain of both cameras was set to 0.0. Before and after each scan, 20 dark frames with matched integration times and gain were captured to allow for dark subtraction of the acquired data. For reference, the optical fibre from the monochromator was also butt coupled to the inspection fibre (TP00895; Avantes) of the calibrated spectrometer and the full spectral scan was repeated.

The acquired data were averaged, dark subtracted and converted to 'spectral response curves' according to Eq. 2.9, taking into account the combined transmission characteristics of the imaging channel and the SRDAs. DN(λ) was taken as the average DN of an ROI drawn over the image of the fibre facet of the optical fibre from the monochromator. The experimentally measured 'spectral response curves' of the two imaging channels were then normalised to the highest peak intensity recorded in the reflectance and fluorescence imaging channel respectively and compared to the predicted spectral response curves. The average RMSE and standard deviation across the spectral response curves in each imaging channel were calculated to evaluate the accuracy of the model.

Experimental Evaluation of Model Accuracy: Endmember Predictions

The accuracy with which the model may be used to predict endmembers for spectral unmixing was experimentally evaluated by comparing modelled and experimentally measured spectra. This section describes the evaluation of the accuracy of the modelled endmembers for spectral unmixing of fluorescent dyes dissolved in PBS. Similar evaluations were repeated for the fluorescent dyes in agarose dye plugs (Sect. 4.2.5) and for the chemically oxy/deoxygenated blood enclosed in glass capillaries (Sect. 4.2.5).

Based on modelling of the fluorescence signals detected by the endoscope, four fluorescent dyes were thought appropriate for demonstration of multiplexed fluorescence imaging; AF647, AF660, AF680 and AF700 (A20006, A20007, A20008, A20110, all NHS ester; Invitrogen). These dyes were selected as the model indicated that strong fluorescence signals would be detected from all four dyes. Whereas supplier provided fluorescence spectra were used in the model, it is widely known that the spectra of fluorescent dye solutions are strongly dependent on the solvent [40]. To obtain accurate reference fluorescence spectra for endmember calculation, the fluorescence spectra of the four dyes dissolved in PBS (10010015; Thermo Fisher) were therefore measured on a plate reader (CLARIOstar; BMG LABTECH). The normalised fluorescence spectra acquired on the plate reader were propagated through the modelled (Sect. 4.2.2) and the experimentally measured (Sect. 4.2.2) spectral response of the fluorescence imaging channel, to model signal detected by the NIR SRDA.

To evaluate the accuracy of the model, equimolar dilutions of the four fluorescent dyes were then endoscopically imaged. 40 μM dilutions of AF647, AF660, AF680 and AF700 were prepared in PBS, and 30 μL of each dye dilution placed in separate wells of a microwell plate (18 well, μ-slide, 81826; ibidi). The well plate was placed under the distal tip of the endoscope, such that the endoscope tip was suspended 10 mm above each well, centrally placed in the endoscopic FOV. Endoscopic imaging of the four dye dilutions were performed following the standard imaging and data pre-processing protocol, described in Sect. 4.2.3. Two replicates of each dye dilution and two control wells containing 30 μL of PBS were imaged. The spectra of the endoscopically imaged fluorescent dyes were extracted from 60 pixel radii circular ROIs, placed over the wells of the well plate using data acquired in the reflectance channel of the endoscope for spatial reference. The experimentally acquired spectra were max-min normalised to remove any systematic noise effects. The accuracy of the modelled fluorescence spectra was subsequently evaluated by comparison with the experimentally acquired ROI spectra.

4.2.3 Imaging Protocol and Data Pre-Processing

Before spectral unmixing of the acquired data, the raw images from the two SRDAs were dark subtracted, demosaicked, co-registered, cropped to an appropriate FOV, honeycomb corrected and the pixel spectra max-min normalised (Fig. 4.6). This section details the image acquisition and pre-processing protocol for the MSI endoscope.

At each imaging session, the exposure times and gains of the visible and NIR SRDAs were first optimised to the sample imaged. For each data acquisition, 10 image frames were simultaneously acquired in the reflectance and fluorescence imaging channel. 10 bit image frames were acquired with the SRDAs, converted and saved as

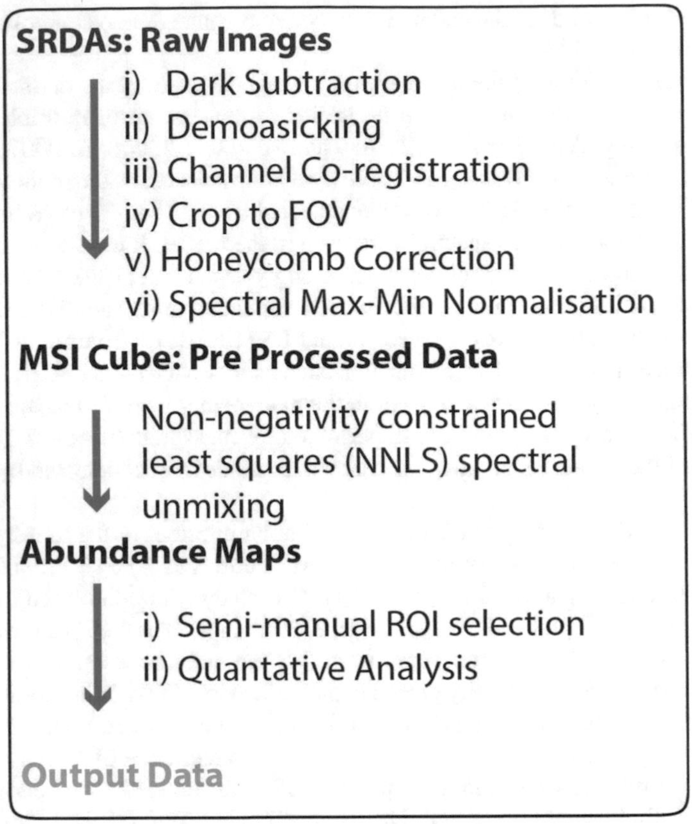

SRDAs: Raw Images

i) Dark Subtraction
ii) Demoasicking
iii) Channel Co-registration
iv) Crop to FOV
v) Honeycomb Correction
vi) Spectral Max-Min Normalisation

MSI Cube: Pre Processed Data

Non-negativity constrained least squares (NNLS) spectral unmixing

Abundance Maps

i) Semi-manual ROI selection
ii) Quantative Analysis

Output Data

Fig. 4.6 Data pre-processing steps performed to acquire abundance maps from the SRDAs in the reflectance and fluorescence imaging channels of the spectral endoscope

16 bit .png files via the LabVIEW® visual interface, using the bit depth information attached to the acquired images.

Prior to data acquisition, 10 dark frames were captured to dark subtract the acquired image data. In video acquisition mode, concurrent 8 bit monochromatic .avi video streams with matched frame rates were recorded from the two SRDAs. Dark videos, containing at least 10 video frames, were captured to dark subtract the acquired data.

The acquired image and video frames were averaged, dark subtracted and demosaicked via linear interpolation of pixels in adjacent spectral macropixels. 16 and 25 spectral band MSI data cubes were thus extracted from the endoscopic reflectance and fluorescence imaging channels respectively. A set of calibration images (re-acquired after each system re-alignment) were used to co-register the two MSI data cubes, to crop the data cubes to an appropriate FOV, and to perform a honeycomb correction; the acquisition of the calibration images and these remaining data pre-processing steps are described in the following sub sections.

Following pre-processing, snapshot and video abundance maps showing the relative position of the blood capillaries and/or the fluorescent dyes were extracted via NNLS spectral unmixing of the MSI data from the reflectance and florescence imaging channels. Pixels with a coefficient of determination below 0.8 were excluded from the abundance maps.

Channel Co-Registration

Due to the different beam paths in the reflectance and fluorescence imaging channels, the images recorded on the visible and NIR SRDAs were not naturally co-registered. The spatial offset and magnification differences of the images (Fig. 4.7a) may arise from the different distances between the objective and the tube lens (the infinity space) in the reflectance and fluorescence imaging channels, as well as the the broadband mirror in the reflectance imaging channel. To overcome the spatial offset of the two MSI data cubes, the MSI data cubes were co-registered by an affine transformation of the NIR MSI data cube (Fig. 4.7b). The affine transformation was defined based on a calibration image of a target with strong salient features (such as a text target) externally illuminated with a broadband light source (OSL2 with OSL2BIR bulb; Thorlabs), such that it could be simultaneously visualised in the reflectance and fluorescence imaging channels.

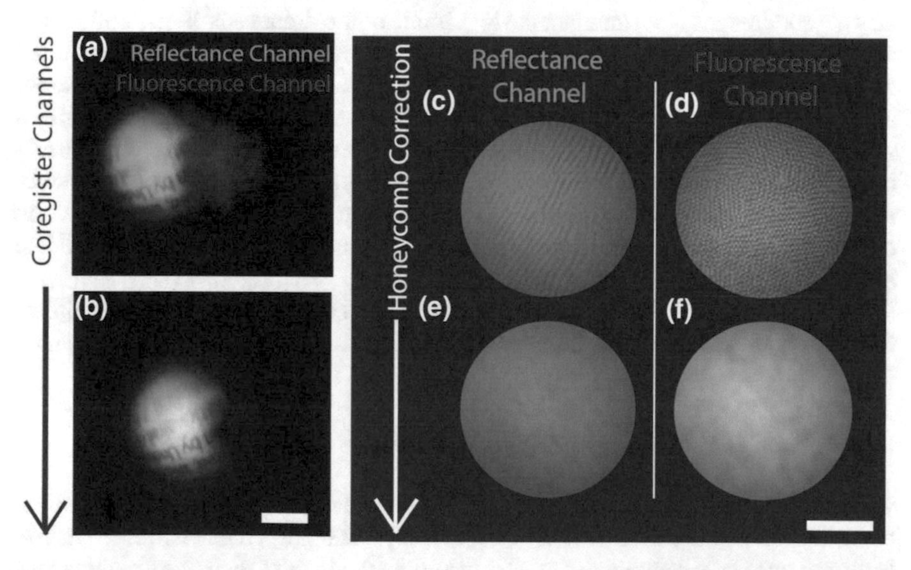

Fig. 4.7 An image of a text target in the reflectance and fluorescence channel before (**a**) and after co-registration (**b**). The co-registered images were then honeycomb corrected to remove the pattern introduced by the individual fibrelets in the CFB of the PolyScope®. Spectral band images of an externally illuminated reflectance target in the reflectance (**c**) and fluorescence (**d**) imaging channels were used to identify the position of the fibrelets in the non-honeycomb corrected images. **e–f** Honeycomb corrected images were then produced via interpolation of the individual fibrelets using Delaunay Triangulation (scale bars = 5 mm)

A set of corresponding reference points in the reflectance and fluorescence channel were extracted via manual identification of salient features in the images using the MATLAB® *cpselect* function. The transformation matrix was subsequently extracted from the set of corresponding image coordinates using the *fitgeotrans* function. A cropping rectangle, defining an appropriate FOV, was manually defined based on the image data from the reflectance channel.

Honeycomb Correction

Images were relayed via a CFB from the proximal to the distal end of the PolyScope®. A honeycomb pattern, in which the individual fibrelets of the CFB appears as bright dots, was therefore visible in the acquired images. The honeycomb pattern is typically removed prior to analysis and data visualisation of image data acquired via CFBs. Here we applied a correction algorithm based on the acquisition of a calibration image, identification of the individual fibre centres via fitting of a Gaussian template and interpolation via Delaunay Triangulation. The correction algorithm was originally presented by Elter et al. [41]. The algorithm was implemented in MATLAB® by Dale Waterhouse (Ph.D. Candidate, University of Cambridge, UK), and has been previously used to correct endoscopic images acquired with a CFB [37].

A reflectance target (Sphere Optics Lambertian White Screen; SG3151-0), externally illuminated with a broadband light source (OSL2 with OSL2BIR bulb; Thorlabs), was imaged to simultaneously visualise the honeycomb pattern in the reflectance (Fig. 4.7c) and fluorescence (Fig. 4.7d) imaging channels. The acquired calibration images were averaged, dark subtracted, co-registered and cropped before applying the honeycomb correction algorithm. The honeycomb correction algorithm was then used to create look up tables (LUTs) for straightforward honeycomb correction of the reflectance and fluorescence imaging channels. The honeycomb correction algorithm was based on the common assumption that the intensity cross section of an individual fibre may be modelled as a 2-D Gaussian function [41]. Candidate fibre centres were initially identified based on their relative brightness ($I_{max} - I_{min}$) in their local neighbourhood ($\mathcal{N}$)—a user defined pixel area surrounding the candidate fibre—provided that the minimum intensity difference exceeds a certain threshold (I_d);

$$I_{max} - I_{min} > I_d, \tag{4.4}$$

which was manually optimised according to the relative intensities of the spectral band images in the reflectance and fluorescence imaging channels. A symmetric Gaussian template, centred on the candidate fibre centre was then fitted to the pixels in the local neighbourhood $\mathcal{N}$. The sum-of-square difference ($s(p_i)$) between the Gaussian template ($T(x, y)$) and the surrounding pixels ($I(x, y)$) was used as a 'score' indicating the likelihood that the identified pixel corresponds to a fibre centre;

$$s(p_i) = \sum_{\mathcal{N}(p_i)} (T(x, y) - I(x, y))^2. \tag{4.5}$$

Starting from the highest ranked fibre (lowest 'score'), the candidate points were added to a centre map, provided that they were at are a minimum distance of d_{min} away from any higher ranked candidate point previously added to the map. A constraint on the minimum separation of the fibre centres was added to prevent double counting of the same fibre. In this manner, fibre centres were continually added to the centre map until all candidate points had either been accepted or rejected. The parameters of the fibre identification algorithm (I_d, $\mathcal{N}$, d_{min}) were manually optimised for the spectral band images in the reflectance and fluorescence imaging channels of the endoscope via visual inspection of the accuracy of the fibrelet identification. The accuracy of the fibrelet identification was evaluated by visually inspecting the fibre centre map superposed on the non-honeycomb corrected spectral band images.

Based on the fibre centre map, a Delaunay Triangulation mesh was calculated via the MATLAB® *delaunay Triangulation* function. Intensity values of intermediate pixels were extracted by weighted addition of the intensities of the closest fibre centres, according to the barycentric coordinates of the pixel in relation to the triangles in the Delaunay Triangulation mesh. Once the procedure was completed, a LUT was defined, which was subsequently used to directly honeycomb correct subsequently acquired spectral band images without further calculation. Interpolation via a Delaunay Triangulation mesh was thought appropriate as fibre centres are connected in a nearest neighbour manner whilst intermediate pixel intensity values are weighted according to their distance from the closest fibre centres.

Following Delaunay Triangulation and interpolation, a Gaussian filter with standard deviation set to half the spectral macropixel dimensions (i.e. 2 pixels for the visible and 2.5 for the NIR SRDA) was applied to the spectral band images. This spatial blurring was applied since no spatial features below the dimensions of the spectral macropixels were to be detected in the spectral band images. It should be noted that the honeycomb pattern appeared visually distinct in the demosaicked spectral band images (Fig. 4.7c, d). Consequently, it can be assumed that the fibrelet images exceeded the spatial dimensions of the spectral macropixels; the filter mosaic pattern of the SRDAs were therefore not taken into account when calculating the LUTs for the honeycomb correction. Since the fibrelets were clearly visible in the spectral band images, the honeycomb correction could be performed after demosaicking the spectral bands. Occasionally misfitting of the fibre candidate points, probably due to slight blurring of the honeycomb pattern due to the filter mosaic, could however be observed in the acquired spectral band images. Although the Gaussian blurring is expected to partially compensate for slight misidentification of the fibrelets, some structure remained even after the honeycomb correction (Fig. 4.7e, f); the performance of the honeycomb correction algorithm was however thought appropriate for the initial performance evaluation of the spectral endoscope performed in this thesis.

4.2.4 Technical Characterisation

After system assembly and establishment of the image acquisition and pre-processing protocol, a technical characterisation of the endoscope was performed to determine the spatial resolution, angular FOV and spectral fidelity of the endoscopic FOV. Analysis code from previously published work by Dale Waterhouse (Ph.D. Candidate, University of Cambridge, UK) [37] was partly used for analysis of the characterisation data. Dale Waterhouse also advised on the design of the technical characterisation experiments.

Spatial Resolution

The spatial resolution of the endoscope was determined at two working distances (WD) by imaging a 1951 United States Air Force (USAF) test chart target (#53-714; Edmund Optics). The spatial resolution was determined by extracting the Michelson contrast from spectral band images of the USAF target in the reflectance channel of the endoscope (Fig. 4.8), acquired and processed according to the standard image acquisition and pre-processing protocol. Since the structure of the fibre bundle was visible in both imaging channels prior to the honeycomb correction, the spatial resolution of the endoscope was assumed to be dependent on the fibre bundle properties, rather than the properties of the SRDAs and should therefore be the same in both imaging channels.

The Michelson contrast (C_m);

$$C_m = \frac{Y_{max} - Y_{min}}{Y_{max} + Y_{min}}, \tag{4.6}$$

is traditionally extracted by fitting a sinusoidal curve over the line pattern of a resolution element of the 1951 USAF target (Fig. 4.8) and extracting the maximum (Y_{max}) and minimum (Y_{min}) intensity from the fitted sine curve. Endoscopic illumination non-uniformities, common to endoscopic imaging [10] arising from effects such as vignetting, could however be observed in the acquired images (Fig. 4.8). This prevented a sinusoidal curve to be straightforwardly fitted to the data. A polynomial baseline correction of the cross sectional intensity profile was therefore performed prior to extracting the Michelson Contrast from a USAF resolution element (Fig. 4.14c).

The images of the USAF test chart target were analysed in MATLAB® according to the following protocol:

Step 1 A resolution element of the USAF test chart target was manually selected.

Step 2 The image was cropped to the area of the selected USAF resolution element and the cropped image rotated such that the line pattern of the USAF resolution element appeared orthogonal to the image x-axis.

Step 3 The cross sectional intensity profile of the USAF resolution element was extracted by averaging the pixel intensities across the y-axis of the cropped image.

Step 4 The function;

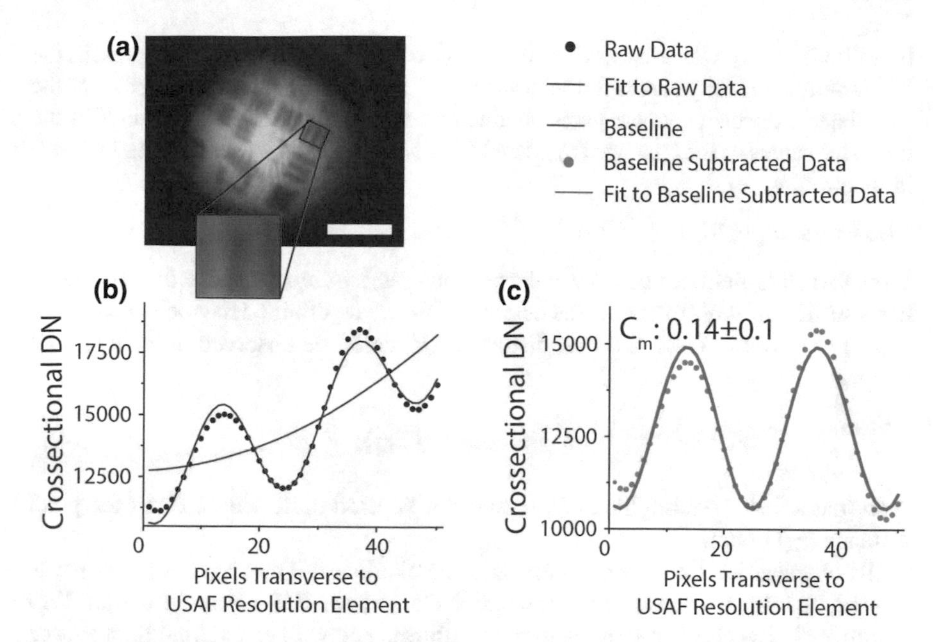

Fig. 4.8 **a** A USAF test chart was imaged to determine the spatial resolution of the endoscope. The experimental method is illustrated with an image of the USAF test chart at a WD of 10 mm, highlighting resolution element 0 in group 2 of the USAF target (1.12 lines/mm) (scale bar = 5 mm). **b** The cross sectional intensity profile across the selected USAF resolution element shows a strong baseline effect due to non-uniformities of the sample illumination. **c** The cross sectional intensity profile was baseline corrected before extracting the Michelson Contrast

$$y = A \sin bx + c + ex^2 + fx + d, \tag{4.7}$$

including the expected sine curve ($A \sin bx + c$) and polynomial baseline ($ex^2 + fx + d$) was fitted to the cross sectional intensity profile of the USAF resolution element.

Step 5 The Michelson contrast was extracted from the cross sectional intensity profile, provided that Eq. 4.7 had a $R^2 \geqslant 0.95$ and that the baseline subtracted data a $R^2 \geqslant 0.80$ when fitting the sine function.

The Michelson contrast was calculated as;

$$\begin{aligned} C_m = \frac{Y_{max} - Y_{min}}{Y_{max} + Y_{min}} &= \frac{(A + B_{avg}) - (-A + B_{avg})}{(A + B_{avg}) + (-A + B_{avg})} \\ &= \frac{A}{B_{avg}}, \end{aligned} \tag{4.8}$$

where A is the amplitude of the fitted sine curve, and B_{avg} the average intensity value across the baseline of the selected USAF resolution element. The average baseline intensity was used as this was considered to most accurately represent the illumination intensity across the selected USAF resolution element. The error of the calculated Michelson contrast was obtained by adding the standard deviation of the intensity values of the baseline (B_{avg}) and the fit error of A (95% confidence interval) in quadrature.

The Endoscopic FOV

To enable wide-field surveillance of the oesophagus, it is important that the endoscope has a sufficient FOV. To extract the angular FOV, we determined the radii of the FOV (r_{FOV}) at a set of WDs. A Barrel distortion (k) could be observed in the acquired images;

$$r_{FOV} = Ar_d(1 + kr_d^2), \tag{4.9}$$

such that a linear scaling factor (A) could not be used to directly convert the pixel distance (r_d) to mm.

To characterise the barrel distortion a checkerboard 1 mm square pattern was printed, and imaged at a set of endoscope WDs between 2.5–15 mm in 2.5 mm WD increments. The checkerboard pattern was illuminated with an external light source (OSL2 with OSL2 bulb; Thorlabs), and the images were acquired and processed according to standard protocol. As in previous work by Waterhouse et al. [37], we identified the position of the centre of the fibre bundle and plotted the radial distance to each vertex in the checkerboard pattern r_d (in pixels), against the true distance r_u (in mm), known from the dimensions of the checkerboard pattern. For each WD, the process was repeated for three spectral bands in the reflectance imaging channel of the endoscope (peak wavelength = 462, 566 and 638 nm) to investigate any potential wavelength dependence of the Barrel distortion. After these characterisation steps, all subsequent imaging was performed at a 10 mm WD, as this gave a sufficient FOV and spatial resolution for the initial imaging performance evaluation of the spectral endoscope.

Spectral Fidelity Over the FOV

To characterise the uniformity of the spectral response over the endoscopic FOV, we defined a 'spectral fidelity metric'. The 'spectral fidelity metric' expresses the MSE deviation of a normalised pixel spectra from that of the average spectral response across the FOV when imaging a uniformly reflective target. The spectral fidelity was experimentally determined by imaging an uniformly reflective target (Lambertian White Screen, SG3151-0; Sphere Optics) according to the standard image acquisition and processing protocol using the endoscope's broadband LED and an external laser line to illuminate the target. To isolate the spectral response of the pixels from amplitude variations, each pixel spectrum was normalised to peak intensity. Following normalisation, the mean pixel spectra across the FOV was calculated and normalised; the MSE deviation of each pixel spectra from the FOV average was used as an indication of the fidelity of the spectral response across the FOV.

Whereas the white light broadband LED could be directly used for spectral fidelity measurements in the reflectance channel, the internal illumination source was modified to achieve broadband illumination of the reflectance target in the fluorescence channel. The white light LED was temporarily replaced with a super-continuum laser source (Super K Extreme EXR-20 supercontinuum laser source and a SuperK Varia tuning unit; NKT Photonics) free space coupled into the input port of the bifurcated fibre; a 750 nm laser line with 40 nm bandwidth was used to illuminate the reflectance target. Since the 'spectral fidelity' across the FOV may depend on a combination of the light source, the SRDA and the optics in the illumination and imaging channel of the endoscope, it is not optimal to characterise the 'spectral fidelity metric' with another NIR light source than the one used for endoscopic imaging. The supercontinuum laser line was however thought to sufficiently well mimic the diode laser line to allow approximation of the overall 'spectral fidelity' across the endoscopic FOV of the fluorescence imaging channel.

4.2.5 Sample Preparation and Imaging

Phantom Preparation and Imaging

To evaluate the spectral imaging performance of the endoscope, we developed a tissue mimicking agarose phantom containing two capillaries of chemically oxy/deoxygenated mouse blood and fluorescent dye inclusions of AF647, AF660 and AF700 dissolved in PBS (Fig. 4.9a). The phantom was designed such that the blood capillaries and the fluorescent dyes could be visualised within the same endoscopic FOV at a 10 mm endoscopic WD (Fig. 4.9b), to demonstrate simultaneous multispectral reflectance and multiplexed fluorescence imaging. The reflectance channel of the

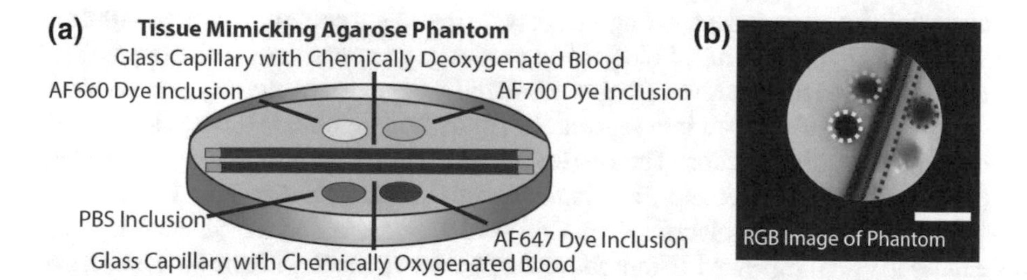

Fig. 4.9 **a** A schematic of the tissue mimicking agarose phantom used to evaluate the endoscopic imaging performance. The phantom contains two capillaries of chemically oxy/deoxygenated mouse blood and fluorescent dye inclusions of AF647, AF660 and AF700 dissolved in PBS. **b** A photo of the tissue mimicking phantom cropped to indicate the FOV of the spectral endoscope at a 10 mm WD; the colour annotations indicates the position of the blood capillaries and the three fluorescent dye inclusions. The phantom was designed such that the blood capillaries and the fluorescent dyes could be visualised within the same endoscopic FOV at a 10 mm WD (scale bar = 5 mm)

endoscope was dedicated to unmixing the oxygenation status of the blood, whilst the fluorescence channel was dedicated to unmixing the three fluorescent dyes.

The phantom base material was prepared as described in Sect. 3.2.3. While liquid, the phantom base material was poured into a 120 mm diameter glass petri dish to form a 5 mm thick agarose layer. Four holes were then made in the agarose slab using a transparent straw with 3 mm internal diameter (391SIPCL; Plastico). The four holes formed a rectangle; the edges of the holes on the short side were touching, whilst the holes on the long side of the rectangle were separated by 6 mm distance. Four 5 mm long pieces of transparent straw, of which one side was sealed with a glue gun (PA6-GF30; Type PX 06; Henkel Pattex Supermatic), were placed in the holes with the glue sealed end facing towards the bottom of the petri dish. Each straw enforced hole was filled with a 80 μM pure fluorescent dye dilution of either AF647, AF660 or AF700 (A20006, A20007, A20110, all NHS ester; Invitrogen) dissolved in PBS (10010015; Thermo Fisher) and one hole filled with a PBS control. Two glass capillaries containing chemically oxy/deoxygenated mouse blood were subsequently placed between the two short sides of the rectangle formed by the fluorescent dye inclusions.

Mouse blood was obtained from 7 female C3H/HeOuJ mice (6 to 10 week old). The mice were euthanised by exposure to raising CO2 concentration. 1 min after the complete cessation of respiratory movements, about 0.5 ml of blood was collected by cardiac puncture from each mouse and placed in an micro-tube tube containing 10 μl of Heparin. The collection of the mouse blood was a non-regulated procedure performed by Dr. Laura Bollepalli (Laboratory Technician, University of Cambridge).

The mouse blood was chemically deoxygenated following a protocol developed by Briley-Sæbø and Bjørnerud [42]; 1.5 mg of Sodium Hydrosulphite (157953-5G-D; Sigma Aldrich) was added and throughly mixed with 0.5 ml of heparnised blood. Successful deoxygenation of the blood was verified with a dissolved oxygen monitor (OxyLite with NX/BF/OT/E probe; Oxford Optronix) on which the deoxygenated blood had a partial oxygen pressure below the detection range of the sensor. Blood was chemically oxygenated by adding 1 μl of 30% hydrogen peroxide (216763–100ML; Sigma-Aldrich) to 0.5 ml of blood. The partial oxygen pressure in the oxygenated blood saturated the dissolved oxygen monitor. Deoxygenated and oxygenated blood were subsequently pulled into separate 100 mm glass capillaries (CV1518; CM Scientific) via capillary action. The capillaries were sealed using capillary tube sealant (02678 Fisherbrand Hemato-Seal Capillary Tube Sealant; Fisher Scientific).

The phantom was placed under the distal tip of the endoscope such that the endoscope was suspended 10 mm above the phantom surface with the fluorescent dye inclusions and blood capillaries centrally placed within the FOV. Snapshot image data were acquired following the standard image acquisition and pre-processing protocol. Three repeat MSI data cubes were acquired, repositioning the phantom between each repeat. A MSI video was also acquired whilst manually moving the phantom under the distal tip of the endoscope.

A blank control phantom containing PBS instead of fluorescent dyes and blood was prepared and imaged using the same protocol. A reflectance target (Sphere Optics

Lambertian White Screen; SG3151-0) was also imaged at a WD of 10 mm following the standard image acquisition and pre-processing protocol.

Reference spectra of the blood capillaries, fluorescent dye inclusions, and phantom base material were acquired with a calibrated bifurcated reflection probe (RP29, Reflection Probe with Linear Leg; Thorlabs) and spectrometer (AvaSpec-ULS2048-USB2-FCDC; Avantes), with the light source leg of the reflection probe connected to the endoscope's broadband LED. The reflectance spectra of the blood containing capillaries were acquired by placing the bifurcated reflection probe flush against the capillaries' surfaces. Reflectance spectra of the phantom background, the PBS inclusion and the combined reflectance/fluorescence spectra of the fluorescent dyes inclusions were acquired in a similar manner (i.e. without laser excitation). A reference spectra of the white light LED source was obtained by placing the reflection probe flush against the reflectance target.

Phantom Data Analysis

Endmember spectra for spectral unmixing of the fluorescent dyes were obtained by propagating the reference fluorescence spectra, previously acquired on the plate-reader, through the modelled spectral response of the fluorescence imaging channel of the endoscope. Endmember spectra for spectral unmixing of the oxygenation status of the blood were calculated based on reference reflectance spectra acquired with the spectrometer. The reflectance spectra of the blood capillaries were converted to optical density (OD) spectra to agree with standard representations of blood oxygenation spectra in the literature;

$$OD = log\left(\frac{I_{white}}{I_{blood}}\right), \tag{4.10}$$

using the reference reflectance spectra from the capillaries (I_{blood}) and the Lambertian white screen (I_{white}). Reference reflectance spectra of the blood capillaries (oxy/deoxygenated blood), the agarose (background) and the reflectance target were used to calculate the endmembers for spectral unmixing of the oxy/deoxygenated blood and the phantom background. Endmembers were obtained by propagating the acquired reference spectra through the modelled spectral response of the reflectance imaging channel of the endoscope prior to OD calculations. The log of the spectral band images acquired in the reflectance channel of the endoscope, divided by the spectral band images of the reflectance target, were calculated to obtain the OD of the MSI reflectance data cube prior to spectral unmixing.

Fluorescent Dye Plugs in an *Ex Vivo* Porcine Model

Endoscopic imaging was performed in an ex vivo porcine model to more closely mimic the real clinical imaging conditions of multiplexed fluorescence imaging within a hollow tube with variable WD. An air-inflated pig stomach and oesophagus (Pig Stomach; Medical Meat Supplies), with internally placed fluorescent dye containing agarose dye plugs, was used as an endoscopic imaging phantom to simulate endoscopic multiplexed fluorescence imaging. As all blood had been drained from the pig stomach and oesophagus, only data from the endoscope's fluorescence channel were analysed.

For preparation of the agarose dye plugs, a 6.0% agar solution (05039–500G; Fluka) was prepared and mixed with equal parts of 160 μM fluorescent dye (AF647, AF660 and AF700) dilutions in PBS. Before solidifying, agarose dye plug droplets were pipetted onto a glass slide to form approximately 5 mm diameter and 1–3 mm thick slabs when solid. The high agarose concentration meant that the dye plugs naturally adhered to surfaces due to their stickiness and high surface tension; they were therefore straightforwardly placed inside the pig oesophagus using a biopsy forcep threaded through the accessory channel of a clinical gastroscope (GIF-H260 gastroscope with Evis Lucera CLV260SL processor; Olympus/KeyMed).

Prior to placing the agarose dye plugs inside the oesophagus, the dye plugs were imaged with the spectral endoscope following the standard image acquisition and pre-processing protocol. Reference fluorescence spectra of the dye plugs were acquired with the calibrated spectrometer and bifurcated reflection probe, following the method described later in Sect. 4.2.5.

Endoscopic imaging was performed by taping the PolyScope® to the side of a clinical gastroscope (GIF-H260 gastroscope with Evis Lucera CLV260SL processor; Olympus/KeyMed) operated by a trained endoscopist (Med. Dr. Massimiliano di Pietro, University of Cambridge). The diameter of the working channel of the diagnostic gastroscope available to us during the imaging was 2.8 mm, whereas the PolyScope® has a atraumatic cap on its distal tip of 3 mm diameter, preventing it from being operating according to the intended babyscope model. The MSI endoscope was therefore taped to the side of the diagnostic gastroscope; this made it difficult to keep the distal tip of the PolyScope® clean and to control the illumination and endoscopic FOV during imaging. Data acquisition was further complicated by the interference between the gastroscope and the MSI endoscope's illumination and detection channels. Light from the gastroscope's internal illumination leaked into the fluorescence channel of the multispectral endoscope, preventing concurrent data acquisition with the two systems.

The gastroscope was instead first used to locate the dye plugs; once the dye plugs had been located, the gastroscope was turned off and data acquisition with the MSI endoscope started. During data acquisition the endoscopist attempted to angle the gastroscope, such that the agarose dye plug would be centrally placed in the endoscopic FOV. Despite efforts to hold the endoscope still, slight endoscopic motion during data acquisition meant that image averaging could not be performed in the subsequent data analysis; apart from these specified differences, the remainder of the standard imaging and pre-processing protocols were followed.

The data from the fluorescence channel of the endoscope were spectrally unmixed using NNLS with endmembers obtained by propagating the reference fluorescence spectra of the agarose dye plugs through the modelled spectral response of the endoscope's fluorescence imaging channel. The accuracy of the modelled spectra was evaluated by comparing the calculated endmembers with ROI spectra of the agarose dye plugs, endoscopically imaged outside the oesophagus.

4.3 Results

4.3.1 *Model of the Endoscope's Spectral Transmission Characteristics*

The spectral transmission characteristics of the endoscope were modelled prior to system assembly by multiplying the transmission spectra of the individual optical components in the endoscopic beam path. The model served a dual purpose; to guide the selection of optical components and fluorescent dyes, and to calculate endmembers for spectral unmixing of the oxy/deoxygenated blood capillaries and fluorescent dyes imaged in this study. After system assembly, the accuracy of the model was evaluated by experimentally measuring the spectral transmission characteristics of the endoscope and by comparing the modelled endmembers to the spectra of those endoscopically measured.

Experimental Evaluation of Model Accuracy: Spectral Transmission Characteristics

The spectral transmission characteristics of the endoscopic illumination channel and the spectral response of the reflectance and fluorescence imaging channels were experimentally measured and compared to the predictions made by the model. A RMSE of 0.26×10^{-3} was obtained between the AUC normalised modelled and experimentally measured transmission spectra of the illumination channel of the endoscope (Fig. 4.10a). Although the RMSE was low and the spectral shape was

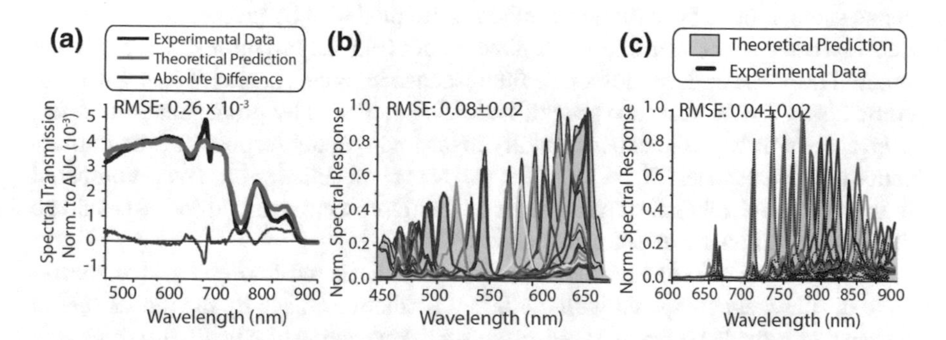

Fig. 4.10 **a** The agreement between the spectral shape of the modelled and experimentally measured spectral transmission of the illumination channel of the spectral endoscope normalized to AUC. The average RMSE of the normalised modelled data in comparison to the experimentally measured data is presented. The difference between the two curves is further highlighted by the plot of the absolute difference between the two curves (red line). A comparison of the modelled and experimentally measured response curves of the spectral bands of the SRDAs in the **b** reflectance (**c**) and fluorescence imaging channel of the endoscope, taking into account the SRDAs, and all other components in the respective imaging channels. The average RMSE of the normalised modelled data in comparison to the experimentally measured data are presented (error; 1 stdv across spectral bands)

found to be comparable, an amplitude offset was observed between the normalised spectra. To avoid the amplitude offset from skewing the spectral comparison, the modelled and experimental data was normalized to AUC. It is a curious coincidence that the amplitude offset between the modelled and experimentally measured spectral transmission occurs at the wavelength of the laser diode (685 nm) included in the endoscope. It may be possible that back reflections of broadband light into the laser cavity during the characterization could have caused stimulated emissions in the laser's diodes gain medium, which would have caused the observed amplitude offset. It should, however be noted the low resolution of the supplier provided transmission spectra used in the model could also have resulted in the observed discrepancies.

Low RMSE was also obtained when comparing the modelled and experimentally measured spectral response of the reflectance (Fig. 4.10b; RMSE: 0.08 ± 0.02) and fluorescence (Fig. 4.10c; RMSE: 0.04 ± 0.02) imaging channels (mean $\pm$ stdv across the spectral bands). Despite the sometimes low spectral resolution of the supplier provided spectra used in the model, the low RMSEs indicate high agreement between the model and the experimental data. This shows that the simple multiplicative model had predictive power.

Experimental Evaluation of Model Accuracy: Endmember Predictions

Fluorescent dyes for imaging demonstrations were selected by modelling the relative strength of the detected fluorescence signals from five fluorescent dyes; AF647, AF660, AF680, AF700 and AF750. Based on modelling of the fluorescence signal strength, four fluorescent dyes (AF647, AF660, AF680 and AF700) were thought appropriate for endoscopic imaging. Experimental measurements of the fluorescent spectra of these four dyes were used to evaluate the accuracy of the modelled fluorescence endmembers. Spectral endmembers were modelled by propagating reference fluorescence spectra, measured on a plate reader (Fig. 4.11a), through the modelled spectral response of the endoscope's fluorescence imaging channel (Fig. 4.11b). For comparison, the fluorescence spectra were also modelled by propagating the reference spectra through the experimentally measured spectral response of the fluorescence imaging channel (Fig. 4.11c). The two sets of modelled spectra were compared to the endoscopically measured spectra of the fluorescent dye dilutions to evaluate the accuracy of the modelled endmembers.

Firstly, the amplitudes of the endoscopically measured fluorescent dye spectra were qualitatively compared with the initial predictions made by the model (based on supplier provided reference spectra). With the exception of AF647, the amplitude of the endoscopically measured fluorescent spectra followed the trend predicted by the model (Fig. 4.11d). In contrast to the other fluorescent dyes, AF647 was mainly excited by light provided by the broadband white light LED rather than the diode laser. The digression from the predicted trend therefore indicates that the relative power provided by the broadband LED and laser diode may have been miscalculated when approximating the spectral distribution of the sample illumination. An

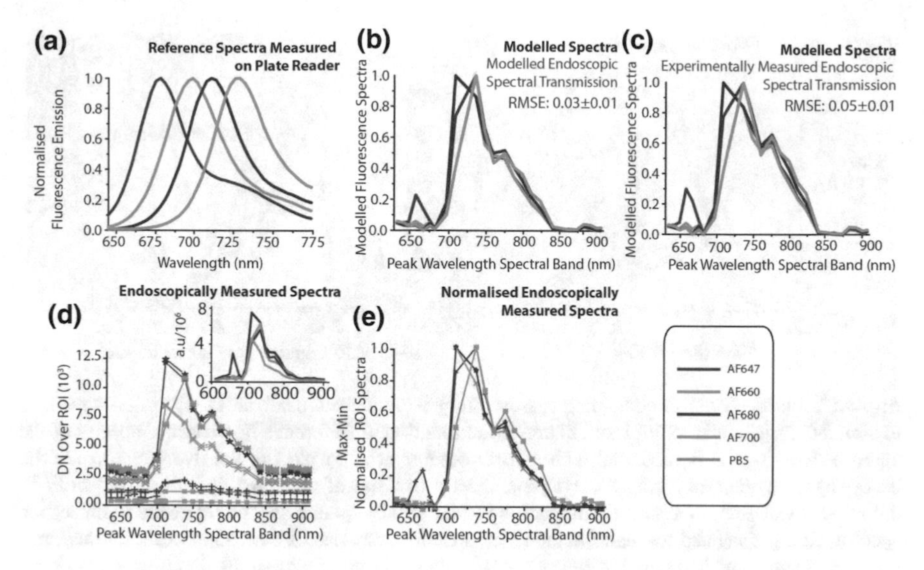

Fig. 4.11 **a** Reference fluorescence spectra of AF647, AF660, AF680 and AF700 dissolved in PBS and measured on a plate reader. **b** Spectral endmembers were modelled by propagating the reference fluorescence spectra through the modelled spectral response of the endoscope's fluorescence imaging channel and the spectral filter band response of the NIR SRDA (sampled by 25 spectral bands). **c** For comparison, the fluorescence spectra were also modelled by propagating the reference spectra through the experimentally measured spectral response of the endoscope's fluorescence imaging channel. To evaluate the accuracy with which the model may be used to calculate spectral endmembers, the two sets of modelled spectra were compared to endoscopically acquired spectra of the fluorescent dye dilutions. **d** The amplitudes of the non-normalised endoscopically measured fluorescence spectra were compared to the fluorescence intensities predicted by the model (see inset). **e** The endoscopically measured fluorescence spectra were then normalised, and compared to the two sets of modelled fluorescence spectra. The RMSEs between the modelled (**b–c**) and experimentally measured (**e**) spectra are indicated in Figure b, c respectively

overestimation of the LED power would also translate to an overestimation of the modelled fluorescence signal strength obtained from the AF647 dye, as the detected fluorescence signal can be expected to be proportional to the light available for fluorescence excitation.

Secondly, the accuracy of the modelled endmembers were evaluated by comparing them to the endoscopically measured fluorescence spectra of the reference dye dilutions (Fig. 4.11e). The endoscopically measured fluorescence spectra showed high agreement with spectra modelled by propagating the reference spectra through the modelled (RMSE: 0.03 ± 0.01), and experimentally measured (RMSE: 0.05 ± 0.01) spectral response of the endoscope's fluorescence imaging channel (mean $\pm$ stdv across the 4 endoscopically imaged dyes). As both methods predict the endmembers with high accuracy, for simplicity only sets of endmember spectra calculated based on the modelled spectral response of the endoscope are presented in the remainder of this chapter.

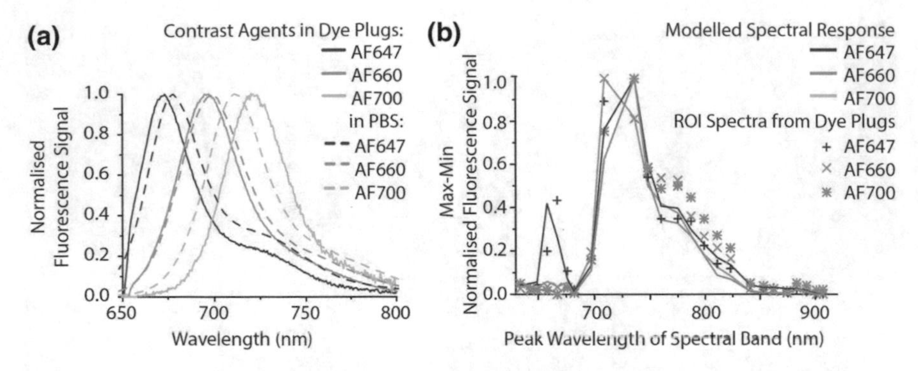

Fig. 4.12 Fluorescent dye containing agarose plugs were placed inside an ex vivo pig oesophagus to simulate endoscopic multiplexed fluorescence imaging. **a** Reference fluorescence spectra of the agarose dye plugs were acquired with a spectrometer prior to placing the dye plugs inside the oesophagus. When comparing the dye plug spectra to those of the same dyes dissolved in PBS, the solvent was seen to drastically impact the fluorescence spectra. **b** The reference fluorescence spectra were propagated through the modelled spectral response of the endoscope's fluorescence imaging channel and compared to the endoscopically acquired spectra of the fluorescent dye plugs, extracted from ROIs placed on the endoscopically acquired MSI data cubes

Experimental data also verified that the normalised endoscopically detected fluorescence signals from AF660 and AF680 were very similar. To reduce spectral misfitting, only AF647, AF660, and AF700 were therefore used in further endoscopic imaging. AF660 was chosen, rather than AF680, despite the latter's higher fluorescence signal, due to the larger availability of this dye.

The modelled spectra were used to spectrally unmix fluorescent dye inclusion of AF647, AF660 and AF680 (dissolved in PBS) placed in a tissue mimicking phantom (Sect. 4.2.5). For imaging of the fluorescent dyes inside an ex vivo pig oesophagus, agarose plugs containing the fluorescent dyes were produced (Sect. 4.2.5). Reference fluorescence spectra of the agarose dye plugs were obtained and propagated through the modelled spectral response of the endoscope's fluorescence imaging channel to calculate endmembers for spectral unmixing of the data. As shown in Fig. 4.12a, dissolving the fluorescent dyes in an agarose-PBS mixture, rather than a pure PBS, drastically impacted the fluorescence spectra of the dyes. The acquired reference spectra of the fluorescent dye plugs were therefore propagated through the model and compared to endoscopically acquired spectra fluorescent dye plug spectra (Fig. 4.12b). The low RMSE (0.07 ± 0.01) between the modelled and endoscopically measured spectra further demonstrates the predictive power of the model.

Spectral endmembers for unmixing of the oxy/deoxygenated blood capillaries included in the tissue mimicking phantom were modelled based on reference reflectance spectra (Fig. 4.13a) propagated through the endoscope's reflectance imaging channel prior to calculating the OD spectra. RMSEs of 0.13 and 0.10 were found when comparing the modelled and endoscopically acquired OD spectra of the oxy/deoxygenated blood capillaries (Fig. 4.13b). These RMSE are relatively high,

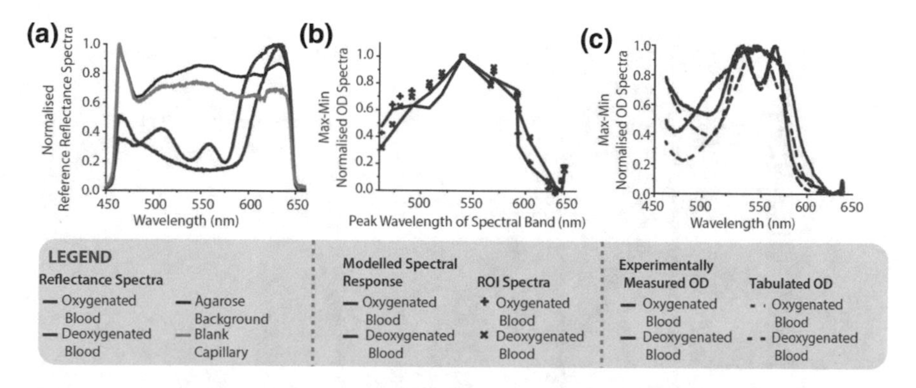

Fig. 4.13 Chemically oxy/deoxygenated blood in glass capillaries were placed on top of a tissue mimicking phantom to demonstrate the spectral unmixing capability in the reflectance channel of the MSI endoscope. **a** Reference reflectance spectra of the glass capillaries containing oxy/deoxygenated blood, a 'blank capillary' (PBS control) and the agarose background were acquired with a spectrometer. **b** Endmember OD spectra for unmixing of the oxy/deoxygenated blood were modelled by propagating the reference reflectance spectra through the modelled spectral response of the endoscope's reflectance imaging channel. The modelled OD spectra were compared with the endoscopically acquired ROI spectra; RMSE of 0.12 and 0.10 were found for the oxy/deoxygenated blood respectively. **c** The OD spectra of the blood was calculated based on the experimentally acquired reference reflectance spectra and compared to tabulated spectra. Tabulated OD spectra were obtained from Bosschaart et al. [43]

but—considering the different experimental procedure of obtaining the reference and the endoscopically acquired spectra—they were however, thought acceptable.

We did also unsuccessfully attempt to calculate endmembers using tabulated oxy/de- oxygenated blood spectra of the effective absorption coefficient, where we assumed the tabulated effective absorption coefficient to be linearly proportional to the OD [43]. The hypothesis was that tabulated spectra could not be used as reference spectra for endmember calculation due to the spectral impact of chemical oxy/deoxygenation of the blood, the imaging geometry and the use of glass capillaries. A discrepancy between the tabulated and experimentally determined OD spectra could indeed be observed in the acquired data (Fig. 4.13c).

4.3.2 *Technical Characterisation*

Spatial Resolution

The Michelson Contrasts for a set of resolution elements of 1951 USAF test chart target imaged at endoscopic WD of 5 and 10 mm were plotted against their corresponding line pairs to determine the spatial resolution of the spectral endoscope (Fig. 4.14). A linear fit was applied to the data and the spatial resolution extracted based on a Michelson cut-off contrast of 1%. A cut-off contrast of 1% was selected,

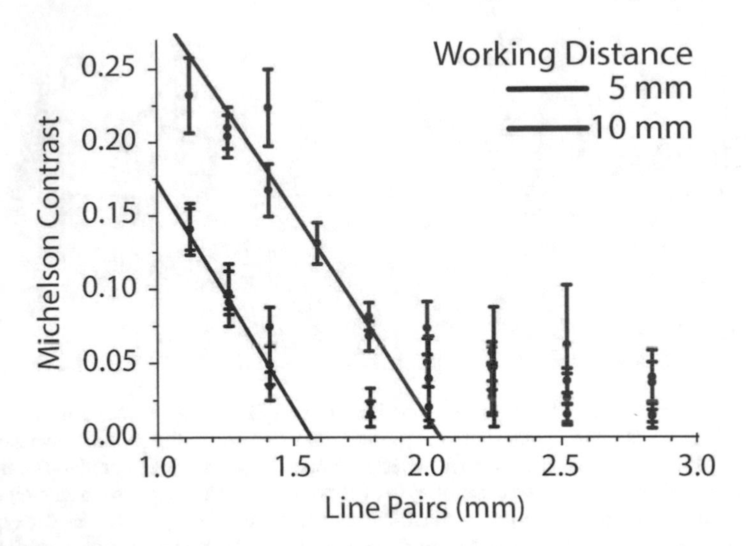

Fig. 4.14 The spatial resolution of the endoscope at 5 and 10 mm WD was determined by imaging an 1951 USAF test chart target. The Michelson Contrast was plotted against the number of line pairs/mm of the resolution elements of the USAF test chart target to determine the spatial resolution of the endoscope

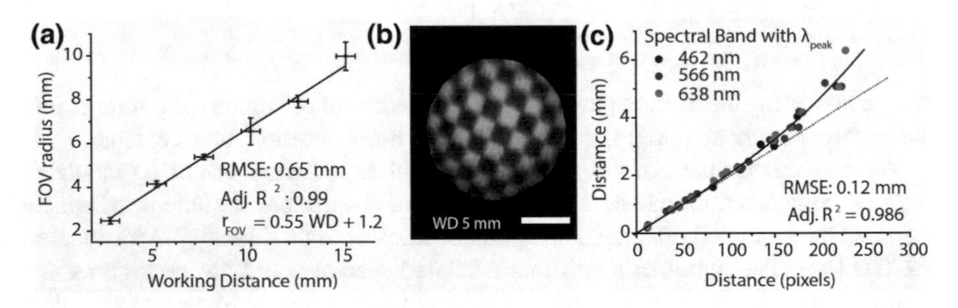

Fig. 4.15 **a** The FOV of the endoscope was measured at a set of WDs, from which an angular FOV of $58 \pm 2°$ was determined. **b** To extract the radius of the FOV at a specific WD, a checkerboard pattern was first imaged to characterise the Barrel distortion (scale bar = 2 mm). **c** The Barrel distortion was extracted by fitting the true radial distance from the centre of the FOV to each vertex in the checkerboard pattern r_d (in pixels) against the true distance r_u (in mm)

as this threshold has previously been shown to represent the minimum resolvable contrast for a wide range of targets independent of their size and luminance [44]. Based on this criteria, the smallest resolvable line pairs/mm were 2.00 ± 0.16 and 1.56 ± 0.17 at 5 and 10 mm WD respectively; this corresponds to spatial resolutions of 0.25 ± 0.04 and 0.3 ± 0.1 mm.

The Endoscopic FOV

The angular FOV of the endoscope was extracted from a linear fit applied to the FOV radii plotted against the endoscopic WDs (Fig. 4.15a). An effective angular

FOV of 58 ± 2°, corresponding to 3.9 ± 0.3 mm and 6.7 ± 0.3 mm FOV radii at 5 and 10 mm WD, was extracted from the gradient of the linear fit (error calculated as the addition of the fit standard error in quadrature). It should however be noted that the FOV was limited by the extent of the illumination rather than the optics of the imaging channel when imaging in air. The FOV of the endoscope was therefore defined as the area within which the diffuse reflectance in all spectral bands exceeded 10% of the maximum reflectance intensity. The diffuse reflectance was determined by acquiring images of the reflectance target (Lambertian White Screen, SG3151-0; Sphere Optics) using the endoscope's internal illumination at WDs from 2.5–15 mm in 2.5 mm increments.

The endoscopically acquired images suffered a slight barrel distortion (Fig. 4.15b), characterised at a set of WDs by imaging a regular checkerboard pattern. For each WD, the Barrel distortion was characterised by plotting the radial distance to each vertex in the checkerboard pattern in pixels (r_d) against the true distance (r_u) in mm, extracted from three spectral bands in the reflectance channel of the endoscope (Fig. 4.15c). In line with expectation, no distinctive wavelength dependence could be observed, and data from the three spectral bands were therefore grouped in the analysis. The radius of the FOV in pixels was then converted into mm taking the effect of the Barrel distortion into account by fitting Eq. 4.9 to the acquired data (Sect. 4.2.4).

Spectral Fidelity Across the FOV

The uniformity of the spectral response across the endoscopic FOV was characterised by imaging a uniformly reflecting target. A 'spectral fidelity metric' was then defined to expresses the mean-square-error (MSE) deviation of pixel spectra from the average spectral response across the FOV. A high 'spectral fidelity metric' was observed in both the reflectance (Fig. 4.16a, average MSE ± stdv: 0.001 ± 0.0009) and fluorescence (Fig. 4.16b, average MSE ± stdv; 0.0003 ± 0.0002) imaging chan-

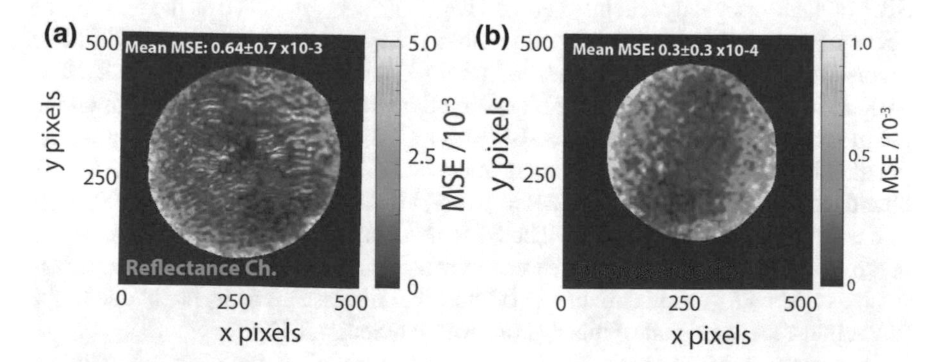

Fig. 4.16 The 'spectral fidelity metric' was defined to express the uniformity of the spectral response across the endoscopic FOV. The MSE deviation of the normalised pixel spectra from the FOV average was used to indication the fidelity of the spectral response across the FOV and produce spectral fidelity maps for the **a** reflectance and **b** fluorescence imaging channels of the endoscope

nels. A small degradation in the spectral fidelity could however be observed towards the edges of the FOV in both imaging channels. The observed degradation in spectral fidelity indicates that the system suffered from slight chromatic aberrations. The non-symmetric nature of the degradation further suggest a slight tilt in one of the optical components in the imaging path. Since the spectral response of pixels at the edges of the FOV deviated from the general spectral response of the endoscope, the spectral unmixing performance may be decreased towards the edges of the FOV. Based on the low MSE we can however assume that such chromatic effects were small, and that the 'spectral fidelity' was generally high across the FOV.

4.3.3 Imaging Performance

A tissue mimicking phantom (Fig. 4.17a) was prepared to evaluate the endoscope's potential for simultaneous spectral reflectance imaging and multiplexed fluorescence imaging. The reflectance imaging channel was here dedicated to unmixing the oxygenation status of chemically oxy/deoxygenated mouse blood, and the fluorescence channel to unmixing the signal from three fluorescent dyes (AF647, AF660 and AF700). After spectral unmixing, a pseudocolour map revealed the oxygenation status of the oxy/deoxygenated blood and identified the fluorescent dyes (Fig. 4.17b), which were not distinguishable in isolated spectral band images (Fig. 4.17c). The blood capillaries (Fig. 4.17d–f) and the fluorescent dye inclusions (Fig.4.17g–j) could each be identified from abundance maps after spectral unmixing via majority decision, based on the endmember abundances extracted from the respective ROIs, and with LS SBRs larger than 3. The spectral unmixing performance of a specific blood capillary/fluorescent dye could be evaluated according to three criteria; the LS score of a ROI drawn over the dye/capillary, the spectral misfitting to the background, and the spectral abundance over the ROI. Fig. 4.17 shows these three parameters. The SBR of a fluorescent dye inclusion/blood capillary was calculated as the ratio of the LS score assigned to the correct endmember within a manually placed ROI to that incorrectly assigned to the background (defined as the area outside the ROI). The SBR and spectral abundance of the blood capillaries and dyes in the tissue mimicking phantom were; oxygenated blood, SBR $= 94 \pm 22$, abundance $= 69 \pm 5\%$; deoxygenated blood, SBR $= 43 \pm 33$, abundance $= 77 \pm 5\%$; AF647, SBR $= 35 \pm 10$, abundance $= 42 \pm 6\%$; AF660, SBR $= 9 \pm 2$, abundance $= 95 \pm 3\%$; AF700, SBR $= 6 \pm 2$, abundance $= 83 \pm 2\%$). The SBR and abundances values are cited as the mean and the range across the three repeat image acquisitions. The imaging performance of AF647 was lower than that of the other fluorescent dyes, likely due to the lower fluorescence signal of this dye (as noted in Sect. 4.3.1).

In video acquisition mode, the oxygenation status of the blood and the fluorescence signal from two of the three fluorescent dyes could be successfully unmixed (Fig. 4.18). The AF647 dye was not detected in video acquisition mode. This may be due to a combination of the lower fluorescence signal from AF647, the inability

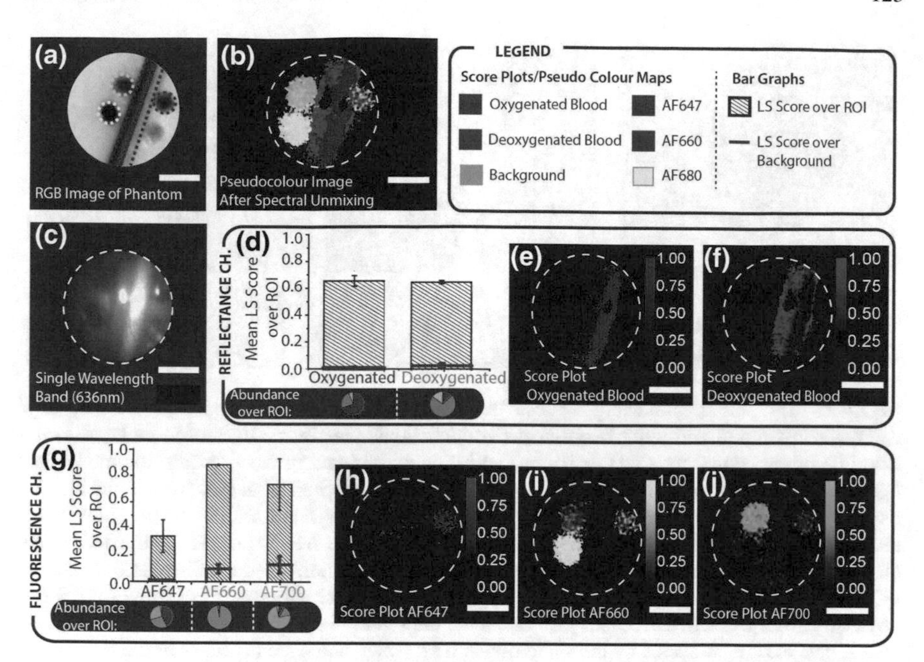

Fig. 4.17 The oxygenation status of chemically oxy/deoxygenated mouse blood and the fluorescent signal from three dyes (AF647, AF660 and AF700) in a tissue mimicking agarose phantom were successfully imaged and unmixed. **a** A standard RGB photo of the tissue mimicking phantom. The phantom was produced to allow the blood capillaries and dye inclusions to be imaged within the same endoscopic FOV. **b** After spectral unmixing a pseudocolour map revealed the position of the dye inclusion and oxygenation status of the blood, which were not visible in single spectral band images (**c**). **d** Quantitative data extracted from the reflectance channel of the endoscope. The LS score (shown in black) was extracted from ROIs placed over the abundance maps of the **e** oxy and **f** deoxygenated blood obtained via spectral unmixing. The error bars indicate the range of LS score over repeat measurements; the LS score of the background is shown in red. **g** Quantitative data from the fluorescence channel were extracted from abundance maps of the **h** AF647, **i** AF660 and **j** AF700 after spectral unmixing (scale bars = 5 mm)

to perform image averaging and the lower DR available when acquiring video rate data.

Endoscopic imaging of fluorescent dye plugs placed inside an ex vivo porcine model (Fig. 4.19a–d) was performed to more closely mimic the clinical conditions of endoscopic multiplexed fluorescence imaging. Two of the three fluorescent dye plugs could be detected inside the pig oesophagus (Fig. 4.19e). AF647 was not detected. Detection inside the pig oesophagus was complicated by the difficulty in controlling the sample illumination and FOV of the endoscope; this may have resulted in lower sample illumination intensities which would have adversely affected the fluorescence signal strength from the dye plugs. AF647 had the lowest fluorescence signal and would therefore be most severely affected by the aforementioned effects.

No quantitative image analysis could be performed on the porcine data since the agarose plugs could not be accurately delineated from the background based on the

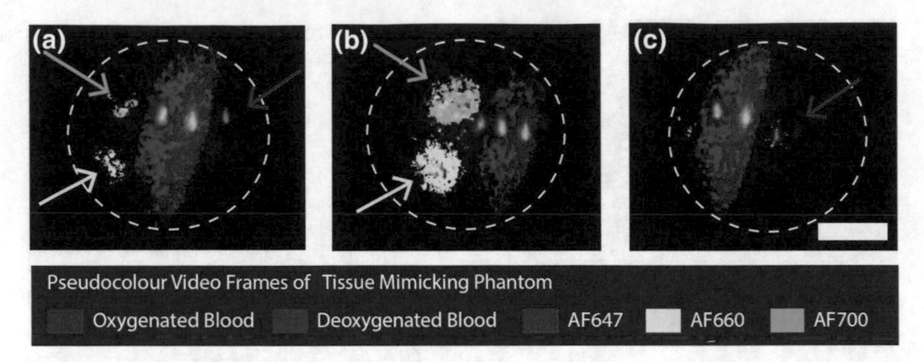

Fig. 4.18 The tissue mimicking agarose phantom, containing two blood capillaries of oxy/deoxygenated blood and three fluorescent dye inclusions, was manually moved underneath the distal tip of the endoscope (WD = 10 mm) whilst acquiring data in video acquisition mode. The figure shows frames from the pseudocolour video produced after spectral unmixing of the blood's oxygenation status and the signal from the three fluorescent dyes. **a** A frame where the blood capillaries were centrally placed within the FOV, **b** a frame where the AF660 and AF700 dye inclusions were centrally placed, and **c** a frame where the AF647 dye inclusion was centrally placed within the FOV. Whereas the blood oxygenation status and the fluorescence signal of AF660 and AF700 can be clearly distinguished in video acquisition mode, the signal from AF647 cannot be detected. The endoscopic FOV is indicated by the white ring and arrows indicate the positions of the fluorescent dye inclusions (scale bar = 5 mm)

acquired images. Additionally, since it was not possible to accurately control the WD of the endoscope inside the ex vivo oesophageal porcine model, the spatial extent of the fluorescence signal could not be accurately sized. As seen in Fig. 4.19e, the fluorescence signals from AF660 and AF700 were however unmixed in FOVs set to include the respective dye plugs; we can therefore conclude that AF660 and AF700 were both detected inside the pig oesophagus.

4.4 Discussion and Conclusions

A bimodal multispectral endoscope for simultaneous spectral reflectance and multiplexed fluorescence imaging was developed by integrating two SRDAs into an endoscopic set-up. The endoscopic imaging performance was demonstrated by simultaneously imaging and spectrally unmixing oxy/deoxygenated blood and three fluorescent dye inclusions of AF647, AF660 and AF700 in a tissue mimicking phantom. We also demonstrated the imaging performance in a more clinically relevant setting by imaging and spectrally unmixing two fluorescent dyes in an ex vivo oesophageal porcine model. In the process of the endoscopy design and assembly we developed general methods to predict and characterise the imaging performance of spectral endoscopes via modelling of the endoscopic spectral transmission characteristics and by defining a 'spectral fidelity metric'.

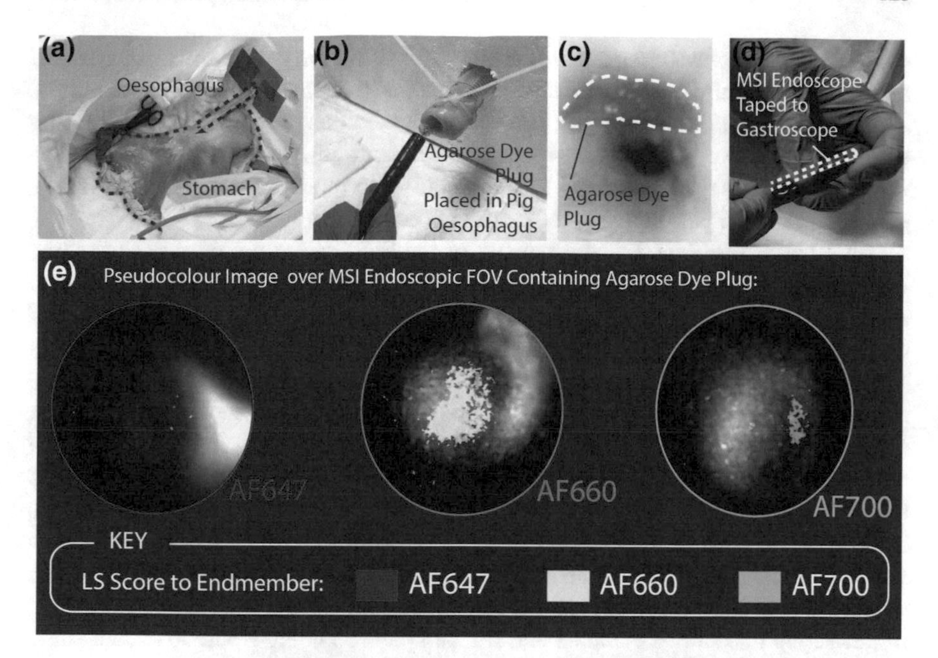

Fig. 4.19 **a** A resected pig oesophagus and stomach was used to mimic clinical endoscopic multiplexed fluorescence imaging. **b** Fluorescent dye containing agarose plugs were placed inside the oesophagus using biopsy forceps threaded through the accessory channel of a clinical gastroscope. **c** Due to the natural stickiness of the agarose dye plugs they adhere to the side of the oesophagus (here visualised inside the oesophagus using the clinical gastroscope endoscope). **d** The MSI endoscope was taped to the side a clinical gastroscope to perform MSI of the agarose dye plugs inside the oesophagus. **e** Pseudocolour abundance maps of FOVs set to contain the three fluorescent dye plugs. Pseudocolour abundance maps were produced via spectral unmixing of the dye signals, overlaid on a summed intensity image of data acquired with the endoscope's reflectance imaging channel. The oesophageal lumen may be observed in the summed intensity images from the endoscope's reflectance channel

By use of two SRDAs, we are able to acquire concurrent visible and NIR MSI data with a compact and robust endoscopy system with potential for low cost manufacture. This may provide a significant cost and size advantage over traditional MSI endoscopes, which tend to rely on multiple optical components for spectral data acquisition, making them costly, large and less robust. In contrast, the bimodal endoscope presented here contains only: one broadband LED, for reflectance imaging; one laser diode, for fluorescence excitation; a dichroic filter; a notch filter, to prevent sensor saturation; and two SRDAs, for concurrent imaging in 41 spectral bands spanning the visible and NIR spectral range. Additionally, it is not inconceivable that the notch filter and the visible/NIR spectral bands may, in the future, be monolithically integrated on the same sensor, creating an even more compact and robust MSI endoscopy system.

The current system does however have a few limitations which need to be addressed. Firstly, fluorescent dyes can currently not be detected at clinically relevant concentrations. In the literature, the typical molarity of fluorescent contrast agents after binding to a specific molecular target is often cited as being within the nanomolar range [29, 45, 46]. All fluorescent dyes in this proof-of-concept imaging study were however prepared at $<40\,\mu$M concentration; yet only two of three fluorescent dyes could be detected in video acquisition mode in the tissue mimicking phantom, or when placed inside the ex vivo porcine model.

As discussed in Sect. 3.4, the sensors of the SRDAs are not optimised for low light applications, such as fluorescence imaging. Depositing the spectral filters on scientific grade sensors could therefore improve the detection efficiency. In this proof of concept imaging study we overcame this challenge by using high molarity fluorescent dye dilutions. The video acquisition detection efficiency of the MSI endoscope could also be improved by modifying the acquisition code to allow video acquisition at the full bit-depth of the cameras. The fluorescence signal from the dyes could also be further increased by employing higher sample illumination intensities. Whereas the combined sample illumination provided by the broadband LED and laser was 3.34 ± 0.02 mW, clinical endoscope operate at about five times this sample intensity [37]. Whereas higher sample illumination intensities may be achieved by a high power halogen lamp and laser, a broadband LED and 685 nm laser diode were used in this set-up to ensure a compact and robust system. The light source choices may however be modified to achieve higher fluorescence sensitivity.

It is also possible that the spectral unmixing performance of the fluorescent dyes may be improved by pooling the spectral data acquired in the reflectance and fluorescence imaging channels. In the current system the reflectance imaging channel was dedicated to unmixing the oxy/deoxygenated blood and the fluorescence channel to unmixing the fluorescent dye signals. By pooling the spectral data, additional spectral contrast could potentially be extracted from the absorption of the fluorescent dyes in the spectral range of the endoscopic reflectance imaging channel. The spectral unmixing performance of the fluorescent dyes may therefore be further improved by exploiting any additional spectral contrast available from the reflectance channel of the endoscope. Currently this was not performed as the fluorescent dyes are mainly excited by the laser line; relatively small spectral differences were therefore observed between the fluorescent dyes in the reflectance channel of the endoscope.

Fluorescence detection within the pig oesophagus was further complicated as the MSI endoscope could not be used according to the intended *babyscope* design. Whereas the PolyScope® is designed to be threaded through the accessory channel of a therapeutic gastroscope, only a diagnostic scope was available during the ex vivo porcine imaging. The smaller size of the accessory channel of the diagnostic scope did not allow the PolyScope® to be operated according to its intended design, the PolyScope® was instead taped to the side of the diagnostic gastroscope. The sub-optimal operation of the MSI endoscope made it difficult to control its WD, the sample illumination and the endoscopic FOV. A therapeutic gastroscope will be required for a fair evaluation of the fluorescence imaging performance of the MSI endoscope when operated inside an ex vivo pig oesophagus.

The oxygenation status of the blood capillaries in the tissue mimicking phantom could be distinguished in both snapshot and video acquisition mode. In contrast to previously reported studies [38, 47] we were however unable to use tabulated reference spectra as endmembers. We hypothesise that tabulated spectra could not be used due to the spectral impact of chemically oxy/deoxygenating the blood, the imaging geometry, and the use of glass capillaries to contain the blood. Endmember spectra were instead calculated by propagating reference spectra, experimentally measured with a spectrometer, through the modelled spectral response of the reflectance imaging channel. As is may not always be possible to acquire accurate reference spectra, it would however be preferable if tabulated spectra could be used as endmembers. To demonstrate the use of tabulated blood oxy/deoxygenation spectra as endmembers, alternative phantom models would however need to be developed. These phantom models should also be designed to produce physiological blood oxygenation levels for a more clinically relevant performance evaluation.

We would also like to expand the work to study the additional tissue contrast that may be obtained from other endogenous chromophores in the visible spectral region, such as oxygenated/reduced NAD, collagen and elastin. In further work, the tissue contrast obtainable from endogenous chromophores could, for example, be explored by imaging excised patient tissue samples.

In the process of endoscope design and assembly, we developed general methods to characterise and predict the imaging performance of spectral endoscopes. Prior to system assembly, we developed a model to predict the spectral transmission characteristics of the endoscope, which was verified via experimental measurements. The model was subsequently used to guide the selection of optical components and successfully calculate endmembers for spectral unmixing. Despite simplistic modelling of coupling losses and fluorescence, the resultant model was shown to have predictive power. The model performance may also be further improved by the use of ray tracing software, which could allow more accurate calculation of coupling losses and chromatic aberrations. A more sophisticated fluorescence model could also be developed to, for example, more accurately take into account the wavelength dependence of the fluorescence absorption.

The spatial resolution, angular FOV and 'spectral fidelity' of the endoscope were determined via a technical characterisation. Whereas the spatial resolution and angular FOV constitutes standard measurements, a 'spectral fidelity metric' was specifically developed to characterise the spectral response uniformity of the endoscope. Although the uniformity of the spectral response is an important performance parameter, we have not found previous examples of attempts to characterise this metric in the literature. In this thesis, we suggest the use of the 'spectral fidelity metric', which expresses the spectral deviations from the average spectral response across the FOV.

Our technical characterisation was somewhat limited, as it was only performed in air. Although this is standard practice in the literature [29, 33, 37], this does not necessarily represent a realistic imaging environment in the GI tract, with its interchangeably air and liquid environment. Ideally the technical characterisation of the endoscope needs to be performed in more realistic imaging environments; this could for example be achieved by performing a technical characterisation of the endoscope with its distal tip submerged in water.

To summarise, a bimodal multispectral endoscope for simultaneous spectral reflectance and multiplexed fluorescence imaging was developed by integrating two SRDAs into an endoscopic set-up. In the process of endoscopy design and assembly, we developed general methods to predict and characterise the imaging performance of spectral endoscopes. Concurrent imaging and unmixing of oxy/deoxygenated blood and the signal from three fluorescent dyes were then demonstrated in a tissue mimicking phantom. Two of the fluorescent dyes were also successfully detected inside an ex vivo oesophageal porcine model. This work demonstrates the potential of SRDAs to enable simultaneous reflectance spectroscopy and multiplexed fluorescence imaging in endoscopy. With further developments, this technology has potential to improve the possibilities for early detection of diseases such as cancer during endoscopic screening of the oesophagus.

References

1. G. Lu et al., Spectral-spatial classification for noninvasive cancer detection using hyperspectral imaging. J. Biomed. Opt. **19**(10), 016004 (2014)
2. L. Gao, R.T. Smith, Optical hyperspectral imaging in microscopy and spectroscopy - a review of data acquisition. J. Biophotonics **8**(6), 441–456 (2015)
3. J. Hoon Lee, T.D. Wang, Molecular endoscopy for targeted imaging in the digestive tract. Lancet. Gastroenterol. Hepatol. **1**(2), 147–155 (2016)
4. R. Krishnamoorthi, P.G. Iyer, Molecular biomarkers added to image-enhanced endoscopic imaging: will they further improve diagnostic accuracy? Best Pract. Res. Clin. Gastroenterol. **29**(4), 561–573 (2015)
5. M.B. Strum, T.D. Wang, Emerging optical methods for surveillance of Barrett's oesophagus. Gut **64**, 1816–1823 (2015)
6. N. Thosani et al., ASGE technology committee systematic review and meta-analysis assessing the ASGE preservation and incorporation of valuable endoscopic innovations thresholds for adopting real-time imaging-assisted endoscopic targeted biopsy during endoscopic surveillance of Barrett's esophagus. Gastrointest. Endosc. **83**(4), 684–698 (2016)
7. R.C. Fitzgerald et al., British society of gastroenterology guidelines on the diagnosis and management of Barrett's oesophagus (2013)
8. K.N. Phoa et al., Multimodality endoscopic eradication for neoplastic Barrett oesophagus: results of an European multicentre study (euro-ii). Gut **65**, 555–562 (2016)
9. M. di Pietro, R.C. Fitzgerald, Research advances in esophageal diseases: bench to bedside. F1000Prime Rep. **5**(44) (2013)
10. B.P. Joshi et al., Multimodal endoscope can quantify wide-field fluorescence detection of Barrett's neoplasia. Endoscopy **48**(2), A1–A13 (2015)
11. C.M. Lee, C.J. Engelbrecht, T.D. Soper, F. Helmchen, E.J. Seibel, Scanning fiber endoscopy with highly flexible, 1 mm catheterscopes for wide-field, full-color imaging. J. Biophotonics **3**(5–6), 385–407 (2010)
12. L. Qiu et al., Multispectral scanning during endoscopy guides biopsy of dysplasia in Barrett's esophagus. Nat. Med. **16**(5), 603–606 (2010)
13. M. di Pietro et al., The combination of autofluorescence endoscopy and molecular biomarkers is a novel diagnostic tool for dysplasia in Barrett's oesophagus. Gut **64**(1), 49–56 (2015)
14. P.B. Garcia-Allende et al., Towards clinically translatable NIR fluorescence molecular guidance for colonoscopy. Biomed. Opt. Express **5**(1), 78–92 (2013)
15. A.K. Shergill, F.A. Farraye, Endoscopic evaluation for colon cancer and dysplasia in patients with inflammatory bowel disease. Tech. Gastrointest. Endosc. **18**(3), 145–151 (2016)

16. F.J.Q. Chedgy, S. Subramaniam, K. Kandiah, S. Thayalasekaran, P. Bhandari Fergus, Acetic acid chromoendoscopy: improving neoplasia detection. World J. Gastroenterol. **22**(25), 5753–5760 (2016)
17. R.T. Kester, N. Bedard, L. Gao, T.S. Tkaczyk, Real-time snapshot hyperspectral imaging endoscope. J. Biomed. Opt. **16**(5), 056005 (2011)
18. Y. Fawzy, S. Lam, H. Zeng, Rapid multispectral endoscopic imaging system for near real-time mapping of the mucosa blood supply in the lung. Biomed. Opt. Express **6**(8), 2980–2990 (2015)
19. T. Saito, H. Yamaguchi, Optical imaging of hemoglobin oxygen saturation using a small number of spectral images for endoscopic application. J. Biomed. Opt. **20**(12), 126011 (2015)
20. S.E. Martinez-Herrera et al., Identification of precancerous lesions by multispectral gastroendoscopy. Signal Image Video Process. **10**(3), 455–462 (2016)
21. S.J. Leavesley et al., Hyperspectral imaging fluorescence excitation scanning for colon cancer detection. J. Biomed. Opt. **21**(10), 104003 (2016)
22. Z. Han, A. Zhang, X. Wang, M.D. Wang, T. Xie, In vivo use of hyperspectral imaging to develop a noncontact endoscopic diagnosis support system for malignant colorectal tumors. J. Biomed. Opt. **21**(1), 016001 (2016)
23. R. Kumashiro et al., Integrated endoscopic system based on optical imaging and hyperspectral data analysis for colorectal cancer detection. Anticancer Res. **36**(8), 3925–3932 (2016)
24. E.L. Bird-Lieberman et al., Molecular imaging using fluorescent lectins permits rapid endoscopic identification of dysplasia in Barrett's esophagus. Nat. Med. **18**(2), 315–321 (2012)
25. J.J. Tjalma et al., Molecular fluorescence endoscopy targeting vascular endothelial growth factor a for improved colorectal polyp detection. J. Nucl. Med. **57**(3), 480–485 (2016)
26. J. Burggraaf et al., Detection of colorectal polyps in humans using an intravenously administered fluorescent peptide targeted against c-Met. Nat. Med. **21**(8), 955–966 (2015)
27. Y. Jiang, Y. Gong, J.H. Rubenstein, T.D. Wang, E.J. Seibel, Toward real-time quantification of fluorescence molecular probes using target/background ratio for guiding biopsy and endoscopic therapy of esophageal neoplasia. J. Med. Imaging **4**(2), 024502 (2017)
28. R. Leitner, M. De Biasio, T. Arnold, C.V. Dinh, M. Loog, Multi-spectral video endoscopy system for the detection of cancerous tissue. Pattern Recognit. Lett. **34**, 85–93 (2013)
29. C. Yang, V.W. Hou, L.Y. Nelson, R.S. Johnston, D. Melville, E.J. Seibel, Scanning fiber endoscope with multiple fluorescence-reflectance imaging channels for guiding biopsy. Proc. SPIE Int. Soc. Opt. Eng. **8936**, 89360R (2014)
30. G. Oh et al., Clinically compatible flexible wide-field multi-color fluorescence endoscopy with a porcine colon. Biomed. Opt. Express **8**(2), 764–775 (2017)
31. X. Gu et al., Image enhancement based on hyperspectral gastroscopic images: a case study. J. Biomed. Opt. **21**(10), 101412 (2016)
32. B. Regeling et al., Hyperspectral imaging using flexible endoscopy for laryngeal cancer detection. Sensors **16**(1288), 1–14 (2016)
33. T.H. Tate, M. Keenan, U. Utzinger, U.J. Black, J.K. Barton, Ultraminiature optical design for multispectral fluorescence imaging endoscopes. J. Biomed. Opt. **22**(2), 036013 (2017)
34. B.P. Joshi, S.J. Miller, C.M. Lee, E.J. Seibel, T.D. Wang, Multispectral endoscopic imaging of colorectal dysplasia in vivo. Gastroenterology **143**(6), 1435–1437 (2012)
35. A. Zeidan, D. Yelin, Spectral imaging using forward-viewing spectrally encoded endoscopy. Biomed. Opt. Express **7**(2), 392–398 (2016)
36. H.T. Lim, V.M. Murukeshan, A four-dimensional snapshot hyperspectral video-endoscope for bio-imaging applications. Sci. Rep. **6**, 24044 (2016)
37. D.J. Waterhouse et al., Design and validation of a near- infrared fluorescence endoscope for detection of early esophageal malignancy. J. Biomed. Opt. **21**(8), 084001 (2016)
38. J. Pichette et al., Intraoperative video-rate hemodynamic response assessment in human cortex using snapshot hyperspectral optical imaging. Neurophotonics **3**(4), 045003 (2016)
39. J. Doke, GRABIT vo 1.0.0.1. Matlab File Exchange (2016), https://uk.mathworks.com/matlabcentral/fileexchange/7173-grabit. Accessed 13 July 2016
40. T. Fisher, Fluorescence Fundamentals: The Molecular Probes Handbook, 11th edn. (2017), https://www.thermofisher.com/ch/en/home/references/molecular-probes-the-handbook.html. Accessed 14 July 2017

41. M. Elter, S. Rupp, C. Winter, Physically motivated reconstruction of fiberscopic images. in *The 18th International Conference on Pattern Recognition* (2006)
42. K. Briley-Sæbø, A. Bjørnerud, Accurate de-oxygenation of ex vivo whole blood using sodium dithionite. Proc. Int. Soc. Mag. Reson. Med. **8** (2000)
43. N. Bosschaart, G.J. Edelman, M.C. Aalders, T.G. van Leeuwen, D.J. Faber, A literature review and novel theoretical approach on the optical properties of whole blood. Laser Med. Sci. **29**(2), 453–479 (2014)
44. D.G. Pelli, P. Bex, Measuring contrast sensitivity. Vision Res. **90**, 10–14 (2013)
45. F. Fantoni et al., Laser line illumination scheme allowing the reduction of background signal and the correction of absorption heterogeneities effects for fluorescence reflectance imaging. J. Biomed. Opt. **20**(10), 106003 (2015)
46. A.V. DSouza, H. Lin, E.R. Henderson, K.S. Samkoe, B.W. Pogue, Review of fluorescence guided surgery systems: identification of key performance capabilities beyond indocyanine green imaging. J. Biomed. Opt. **21**(8), 080901 (2016)
47. H. Li et al., Snapshot hyperspectral retinal imaging using compact spectral resolving detector array. J. Biophotonics **10**(6–7), 830–839 (2016)

Chapter 5
Conclusions and Outlook

This thesis demonstrates the potential of hyper and multispectral cameras based on commercial SRDA technology for multiplexed biomedical fluorescence imaging. Following a detailed calibration of commercial SRDAs, their potential in two biomedical imaging applications were studied; wide-field fHSI and MSI endoscopy. In the process of system integration of the sensors, the calibration methods, instrumentation challenges and data analysis protocols for effective use of SRDAs in biomedical imaging applications were explored.

In Chap. 2, a standard camera calibration was first performed to verify the supplier provided QE data. We also found that the spectral response of the SRDAs are highly dependent on the AOI of the light. Yet, in the literature, there are only relatively limited experimental studies of the angular sensitivity of the spectral response of SRDAs. Following validation of the supplier provided QE data, a monochromator-integrating sphere set-up incorporating a variable F/# objective was therefore developed to study the angular dependence of the sensors' spectral response. In accordance with the theory of light interactions within Fabry–Pérot cavities [1–3], our data show that imaging at a lower F/# causes a broadening of the FWHM and a blue shift of the peak wavelength of the SRDAs' spectral filters. We also noted that the baseline of the spectral response of the Imec NIR snapshot and linescan SRDAs increased drastically when imaging at lower F/#s. Our results highlight that the angular sensitivity of SRDAs should be carefully consider when integrating these sensors into optical systems. These results were used to guide the selection of accessory optics for further work with the sensors.

With additional refinements of the F/# characterisation set-up and data analysis methods, it may also be possible to obtain experimental data to perform full system-level calibrations of each sensor at a set of F/#s. The construction of calibration matrices representing the pixel level spectral response of the sensors could for example be used to correct for spectral non-uniformities across the FOV and to more accurately calculate endmembers for spectral unmixing. Although the current

A. S. Luthman, *Spectrally Resolved Detector Arrays for Multiplexed Biomedical Fluorescence Imaging*, Springer Theses,
https://doi.org/10.1007/978-3-319-98255-7_5

experimental dataset was sufficient to study the general F/# dependence, and to guide the selection of accessory optics, the sensor calibration could be further developed to provide a full system level calibration. Improvements to the experimental set-up would, for example, include characterising and correcting for any illumination non-uniformities at each wavelength of the spectral scan. Following successful construction of a calibration matrix, the pixel level spectral response corrections would also need to be efficiently incorporated in the software used for data pre-processing and spectral unmixing.

As detailed in Chap. 3, after the sensor calibration we integrated a linescan SRDA in a wide-field fHSI imaging set-up and used two snapshot SRDAs to realise multi-spectral endoscopy. With these two system we demonstrated multiplexed reflectance and fluorescence imaging with SRDA technology in solution, in tissue mimicking and ex vivo phantoms and in vivo in a mouse model. Using the wide-field fHSI system we were able to accurately resolve seven fluorescent dyes in solution. We also demonstrated high spectral unmixing precision, signal linearity with dye concentration, at depth in tissue mimicking phantoms and delineation of four fluorescent dyes in vivo. Although this system has intrinsic value as a wide-field fluorescence imaging platform, it also allowed us to explore the implementation challenges associated with SRDA technology without the added complexities associated with a full intraoperative/endoscopic integration.

The wide-field fHSI system is mainly limited by its low fluorescence detection efficiency. This problem could be partly solved by increasing the obtainable fluorescence signal from the dyes by use of higher intensity sample illumination, and alternative illumination geometries. To mimic the illumination geometries often encountered in the clinic, we chose to perform reflectance based imaging. Alternative illumination-detection geometries could, however, improve the imaging performance of the wide-field fHSI system [4]. Orthogonal illumination-detection geometries could, for example, prevent camera saturation at high illumination intensities and limit specular reflections and glare. Further exploration of alternative illumination and detection geometries could also help achieve quantitative depth resolved fluorescence imaging of extended 3D objects [5]. Depth resolved fluorescence imaging would also require corrections for variable WDs and modelling of light propagation in tissue [5–8].

Following the successful demonstration of multiplexed fluorescence imaging in a wide-field set-up, we proceeded to endoscopic integration of two snapshot SRDAs to realise a MSI endoscope (Chap. 4). The bimodal multispectral endoscope was designed to enable concurrent spectral reflectance imaging of intrinsic chromophores of oxy and deoxyhaemoglobin in the visible, and multiplexed imaging of extrinsic fluorescent contrast agents in the NIR. In the process of endoscope design and assembly, we developed and experimentally verified a model of the spectral transmission characteristics of the endoscope. We also developed and adapted standard endoscopy characterisation methods for characterisation of the spectral endoscope. The imaging performance of the endoscope was subsequently demonstrated by simultaneous imaging and spectral unmixing of oxy/deoxygenated blood and three fluorescent dye inclusions in a tissue mimicking phantom. Two fluorescent dyes were also detected inside an ex vivo oesophageal porcine model. Further work is however required before clinical application of the MSI endoscope. The preliminary imaging data presented here do however show that spectral endoscopy based on SRDA technology has the potential to enable multiplexed spectral reflectance and fluorescence imaging in the clinical setting.

A common challenge for multiplexed fluorescence imaging with SRDAs, highlighted by both the wide-field and endoscopic sensor integrations, is the low fluorescence detection efficiencies of the sensors. The commercial SRDAs used here are not developed for low light applications, such as fluorescence imaging, and do not have the sensitivities required to detect clinically relevant fluorescence signals in the current system integrations. Achieving the nanomolar fluorescence detection efficiencies required for clinical molecular imaging would require a 100-fold improvement in the fluorescence detection efficiency. Whereas the detection efficiency of the current system may be partly improved by optimising the fluorescence excitation, and by minimising coupling losses in the optical systems, further sensor improvements will most likely be required.

The detection efficiencies of the SRDAs could probably be improved by an additional factor of 10 by optimising the spectral filter mosaic pattern and by depositing the spectral filters on scientific grade charge-couple-device (CCD) sensors. The detection efficiency of a fluorescent dye may, for example, be improved by increasing the bandpass of its spectrally matched filter in the filter array. Increasing the bandpass of a certain spectral band would, however, lead to an associated decrease in the spectral resolution. The development of custom filter arrays should therefore be paired with further literature exploration and modeling to optimally match the filter array design to the image application. The linescan HSI system, with its high spectral and spatial resolution, could also be used for initial experimental studies to establish the optimal spectral bands for a specific imaging application.

Custom HSI/MSI cameras with monolithically integrated spectral filters (as developed by Imec) are also expensive, but cheaper custom SRDA camera systems may be realisable via deposition of the spectral filters on a glass substrate placed in front of an image sensor. As identified by our characterisation of the SILIOS MSI camera, this can incur additional noise due to optical interference between the glass substrate and the image sensor. Due to the strong need for cost effective, custom filter designs deposited on high grade scientific grade CCD sensors, methods to further characterise and correct for additional artifacts arising from glass deposition of spectral filters should, however, be developed.

Depositing the spectral filters on a scientific grade sensor could be a relatively straight-forward approach to improve the fluorescence detection efficiency of the SRDAs. However, whereas the CMOS sensor of the snapshot SRDAs used in this thesis weigh less than 50 g and cost less than 500 £, a high performance cooled electron-multiplying charge-couple-device (EMCCD) camera can weigh up to 5 kg and cost towards 50,000 £. The cost and weight of high performance CCD cameras would therefore have adverse effects on the system cost and compactness.

For endoscopy, a more realizable solution would therefore be to integrate the SRDAs on the endoscope tip to side-step the transmission losses and spatial down-sampling introduced when imaging through a fibre bundle. It should be noted that, whereas we integrated two SRDAs at the back end of a modular clinical endoscope, the image sensors of flexible clinical endoscopes are typically integrated on its distal tip [9]. The acquisition of spectral data has, however, traditionally required bulky and sensitive instrumentation, which has prevented the integration of spectral image sensors on tip. With further developments of the lithography techniques used to realise SRDAs, it could become possible to integrate SRDAs on the endoscope tip. MSI data acquisition on tip could improve the image quality and detection efficiencies by reducing the coupling losses and complexity of the endoscopic optical system. Hence, whereas Chap. 4 demonstrates the potential of SRDAs for multiplexed endoscopic imaging, future chip-on-tip implementations have the potential to further improve the imaging performance.

In our work, we observed that the fluorescence signals of fluorescent dyes in dilution are strongly dependent on the solvent. When imaging in vivo we further expect the recorded fluorescence signal to be modulated by the optical properties of tissue. As discussed in Sect. 3.4, these effects are expected to be especially problematic when imaging extended 3D objects at depth [6, 7]. Environment dependent changes of the fluorescence signal can be problematic when extracting endmembers for supervised spectral unmixing. Here, we approached the environment dependence of the fluorescence signal differently for the wide-field fHSI system and the MSI endoscope. A separate 'reference' sample of the fluorescent dyes was imaged with the wide-field fHSI system to extract endmembers for unmixing of the fHSI data. Endmembers for unmixing the endoscopically acquired MSI data were instead modelled based on propagating separately recorded reference spectra through the modelled spectral response of the endoscope's imaging channels. The successful implementation of both approaches indicate some flexibility in the extraction of endmembers for spectral unmixing.

For future clinical implementation, there may be merit in also exploring unsupervised unmixing methods to allow unmixing of extrinsic fluorescence signals on a wide variety of tissue backgrounds without the prior acquisition or knowledge of reference endmembers. In clinical applications, data pre-processing and spectral unmixing also need to be performed in real-time. Further work is therefore needed to optimise the data processing algorithms and transfer them to parallel processing units, such as GPUs or FPGAs, in order to achieve real-time spectral classification.

In the future, one might also imagine that the spectral dimension of the data could be further exploited to perform real-time WD corrections of the endoscopically acquired data. Akin to the endoscopic WD correction method presented by Joshi et al. [10], one may acquire a wide-field reflectance image for ratiometric correction of the fluorescence data. Instead of subsequently acquiring the reflectance and fluorescence images, such as in the work by Joshi et al., one may exploit the spectral data dimension of the SRDA to simultaneously acquire the wide-field reflectance correction image and spectral reflectance and fluorescence data. One spectral band, with a matched narrowband light source, could for example be used to acquire the correction image, whilst simultaneously acquiring reflectance and fluorescence spectral data in the remaining bands.

To enable clinical multiplexed fluorescence imaging, instrument and software advances also need to be paired with further developments of targeted fluorescent contrast agents. Although numerous targeted fluorescent contrast agents have been developed within the last 10 years, clinically multiplexed fluorescence imaging is currently held back by the lengthy and costly process of obtaining regulatory approval for novel imaging probes. The diagnostic use of multiplexed biomedical reflectance and fluorescence imaging also needs to be further verified. Whereas this thesis focused on technology integration, characterisation and imaging performance evaluation of SRDA technology, the potential of the technology to improve the diagnostic performance in real clinical imaging applications should also be studied. This motivated us to design our multispectral endoscope for straightforward translation into clinical imaging applications, by assembling it around a CE marked modular clinical endoscope. In further work the diagnostic potential of the MSI endoscope may therefore be straightforwardly compared to standard clinical procedures.

To summarise, this thesis demonstrates the potential of MSI/HSI cameras based on SRDA technology for multiplexed biomedical imaging of intrinsic chromophores and extrinsic fluorescence contrast. The technology integration, characterisation and imaging performance evaluation of commercial SRDA technology were evaluated in a wide-field and an endoscopic imaging set-up. With further technology developments, we showed that SRDAs have the potential to acquire hyper/multispectral data in a robust, compact, and potentially very cost effective fashion. This approach therefore holds promise to be able to increase the tissue contrast in biomedical and clinical imaging.

References

1. H.A. Macleod, Chapter 6: Edge filters and Chapter 7: Band-pass filters, *Thin Film Optical Filters*, 3rd edn. (IoP, Bristol, 2000)
2. M. Jayapala et al., Monolithic integration of flexible spectral filters with CMOS image sensors at wafer level for low cost hyperspectral imaging in international image sensor workshop, in *Snowbird* (2013)
3. P. Agrawal et al., Characterization of VNIR hyperspectral sensors with monolithically integrated optical filters, in *Proceedings of IS&T International Symposium on Electronic Imaging, 2016* (2016)
4. T. Sawyer, A.S. Luthman, S.E. Bohndiek, Evaluation of illumination system uniformity for wide-field biomedical hyperspectral imaging. J. Opt. **19**(4), 045301 (2017)
5. A.J. Chaudhari et al., Hyperspectral and multispectral bioluminescence optical tomography for small animal imaging. Phys. Med. Biol. **50**, 5421–5441 (2005)
6. P.A. Valdes et al., Quantitative, spectrally-resolved intraoperative fluorescence imaging. Sci. Rep. **2**(798) (2012)
7. M. Jermyn et al., Macroscopic-imaging technique for subsurface quantification of near-infrared markers during surgery. J. Biomed. Opt. **20**(3), 036014 (2015)

8. S. Tzoumas, N.C. Deliolanis, S. Morscher, V. Ntziachristos, Unmixing molecular agents from absorbing tissue in multispectral optoacoustic tomography. IEEE Trans. Med. Imaging **33**(1), 48–60 (2014)
9. ASGE, High-definition and high-magnification endoscopes. Gastrointest. Endosc. **80**(6), 919–927 (2014)
10. B.P. Joshi et al., Multimodal endoscope can quantify wide-field fluorescence detection of Barrett's neoplasia. Endoscopy **48**(2), A1–A13 (2015)

Appendix
Components Tested in Endoscope Model

When designing a spectral endoscope, the spectral transmission characteristics of each of the optical components in the beam path need to be considered. To guide the selection of optical components for endoscopic set-up, a MATLAB® model was developed to predict the spectral characteristics of the full endoscopic illumination and imaging channel prior to system assembly. The following components were tested in the model:

- **LEDs**: T7358, T7359 (Prizmatix)
- **Lasers**: LP660, LP785, M740F2, M780F2 (Thorlabs)
- **Longpass Filters**: AT465, CT485lp (Chroma), LPEO475OD2, LPEO475OD4 (Edmund Optics)
- **Shortpass Filters**: FESH0650 (Thorlabs), FES0600 (Chroma), FF01-650 (Semrock)
- **Dichroic Filters**: FF652-Di01, FF635-Di01 (Semrock), ET655 (Chroma), DMLP650 (Thorlabs)
- **Notch Filter**: ZET685NF, NF03-685E, ZET785 (Chroma), NF785-33 (Thorlabs)

The remaining optical components had relatively flat spectral profiles; due to their limited effect of the overall transmission characteristics of the endoscope, these components were kept constant in the model.

A. S. Luthman, *Spectrally Resolved Detector Arrays for Multiplexed Biomedical Fluorescence Imaging*, Springer Theses,
https://doi.org/10.1007/978-3-319-98255-7